Rules Of Many Knightly Exercises

Rules Of Many Knightly Exercises

Collected by Captain Federico Ghislieri (1587)

Mike Smallridge
Michael Chidester

HEMA Bookshelf

The following work is based upon the copy of Federico Fabio Ghislieri's *Regole di molti cavagliereschi essercitii* (1587) held by the Bibliothèque nationale de France. ark:/12148/bpt6k1518476n

Published by HEMA Bookshelf, LLC.

47 High Street #443
Medford, MA 02155
United States of America
www.hemabookshelf.com

Version 1.0, 2026.

Hardcover ISBN: 978-1-953683-61-8
Softcover ISBN: 978-1-953683-62-5

Printed by Ingram Lightning Source.

Table of Contents

Practice

Siege-Craft Analogy

The Barriers

Horse

Appendices

Editor's Foreword

Federico Fabio Ghislieri[1] was an Italian officer and engineer of the Counter-Reformation, active from 1580-1621. In 1587 he had a book printed for the benefit of young noblemen describing the rules (or principles) of combat.

This translation is the result of several years of effort researching and interpreting Federico's work. I first came across its gorgeous illustrations in Sydney Anglo's *The Martial Arts of Renaissance Europe* (2000). For a long time I was perplexed by the pleasant, complicated drawings. Thanks to modern digitalisation, we can now understand how Ghislieri worked from first principles to distil sword and dagger combat down to just four techniques, chosen based on how the opponent holds their sword.

I've tried to make the text easily readable for a modern audience, and so translated the fencing jargon. If you have any doubts about a particular choice (or want inspiration for your next coffee order), a transcription of the original Italian is included at the back of the book. Cross-reference between the two using the page (p.) and section (§) references to the left of the text.

If your main intention is to apply the ideas in this book to modern historical fencing, I'd recommend skipping ahead to p. 60 §4 of the text. This is where Ghislieri lays out the fundamental problem in fencing, previous attempts to solve it, and begins explaining his own solutions for rapier and dagger combat.

I've also made the editorial decision to move a few sections of text. For the most part these are minor changes to improve the books layout. However, the original text has a long intermission between the theory and practive halves of the book. While that extended analogy between fencing and siege-craft is interesting, it also makes it harder to follow Ghislieri's fencing techniques. It can now be found after the chapter on fencing lefties.

Many Knightly Exercises was composed near the beginning of Ghislieri's career, after only a few campaigns with Hapsburg forces in Portugal, Flanders and the Rhine. Three major themes run through the text:

1. *Military Theory* —The author spends considerable effort linking fencing theory to military engineering. Ghislieri's martial approach to rapier and dagger fencing uses

1 I've chosen to use 'Ghislieri' (the plural form of Federico's family name), like continental academics. The original text used the singular form 'Ghisliero', which has been followed in English until now.

a measured aggression to overwhelm the opponent as safely as possible. It draws on geometry and science to explain the principles of the system, and illustrates them with gorgeous pen and ink drawings.

2. *Pre-Newtonian Thinking* — Ghislieri is keen to show his knowledge of Aristotelian physics and classical thought. During this period of Catholic counter-reformation, Scholastic thinking was undergoing a revival due to the influence of the Jesuit movement. The text is a fascinating glimpse into a bright scientific mind, before Descartes, Locke, or Newton.
3. *Interconnected Ideas* — The fencing system presented in the book has its idiosyncrasies, but it's a remarkable example of the hugely influential approach of Silvio Piccolomini. Capoferro also lauded Piccolomini, and it's probable that he helped in the publication of that book through a learned society. Later in his life, Ghislieri also served alongside Giganti and probably Gaiani. The text eschews the 'Bolognese' fencing tradition in favour of an Agrippan approach, but never-the-less shows familiarity with the ideas and vocabulary of earlier sources. In some sections he even takes from other books for his collection!

Unlike many swordsmen of the past, whose lives are lost to time apart from a few lines of lesson notes, Ghislieri followed his father into a long and successful military career. Fighting for various Catholic powers across Europe and the Mediterranean, against Protestants and Ottomans, Ghislieri corresponded with the Pope, the Holy Roman Emperor, and hosted Galileo. For more information on Ghislieri's life, and the contemporary fencing environment, see the timeline in the appendices. Despite being seriously wounded in a Hungarian siege, Federico's service culminated in a promotion to the nobility in Savoy. His life embodied the early modern military revolution, in which states employed logistics and engineering experts to lead their ever-growing, ever-professionalising armies needed to pursue the political ambitions of their rulers.

Many Knightly Exercises is an odd book in several ways — printed with blank spaces to be illustrated later, and at least two alternate dedications. The work has gone over-looked by posterity, eclipsed until now by contemporaries who published later in life, such as Fabris (1601), Giganti (1606 and 1608), and Capoferro (1610). The production of the book probably served to further Ghislieri's career, ingratiating him with noble cliques commanding Imperial forces, but it's hard to judge it's success at the time. Ghislieri had a long and successful military career, but through the Papal States and then Savoy.

I hope that the reader finds the text as interesting and informative as I have.

Mike Smallridge, Bristol, 2026

Translation Notes

The author refers to himself as Federico Ghisliero, which is how English authors have tended to refer to him. In modern Italian, his name would be given as Ghislieri, the plural form. I've adopted the latter for the translation, following the example of modern scholarship (see the historiography appendix for more information). If it needs further justification, the Spanish ambassador at the time referred to him as *"Federigo Ghisilieri"*, which a letter from a colleague on campaign referred to him as *"Il Cap.^{no} Fidirigo Ghiselieri Sergenti Magg.^{ior}"*.[1]

The text tends to use "retta" to mean the shortest–line between two points, and "diritta" or "manca" when talking about sides. The section on left–handers uses capitalised Manca and Diritta to make emphasise that these are the two fencers, our Righty and Lefty.

The author had also a tendency to use long form numbered lists, which can get confusing when those numbers are ambiguous between movements, concepts, and positions. Footnotes have been used to clarify, along with interpretative notes in the translation, [clearly marked] with square brackets.

Ghislieri uses four numbered approaches when engaging an opponent, informed by how the opponent is holding their sword. This is made more confusing since the opponent's sword position is often referred to according to a different scheme of four numbered guards, stemming from Agrippa's 1553 fencing book.

Agrippa's system starts with the first guard, which has the sword hand totally pronated (that is with the palm facing outwards and the thumb pointing down), as when drawing a sword from its scabbard. The guards then rotate around (clockwise for a right-handed fencer, anti-clockwise for a left-handed fencer) until the sword hand is on the 'inside' of the front knee. I've referred to these 'Agrippan' positions as 1st, 2nd, 3rd, and 4th for the sake of clarity. Here are those positions illustrated in Agrippa:[2]

---

1 17th of January 1616– Federico Ghisliero hosts Galileo Galilei in Rome – Antonio Querengo to Alessandro D'Este, 20 January 1616, *Opere*, vol. 12, pp. 226-7.

23rd of July 1601 – 1 Archivio di Stato di Firenze, Mediceo del Principato, Vol. 5155, p. 27r.

2 The illustrations are taken from https://hdl.handle.net/2027/gri.ark:/13960/t2c83659q and are in the public domain.

1st

2nd

3rd

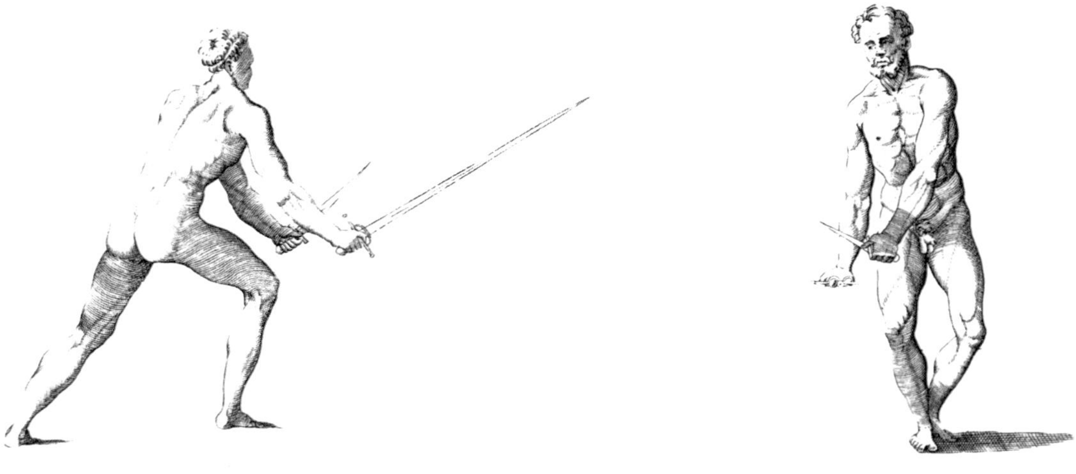

4th

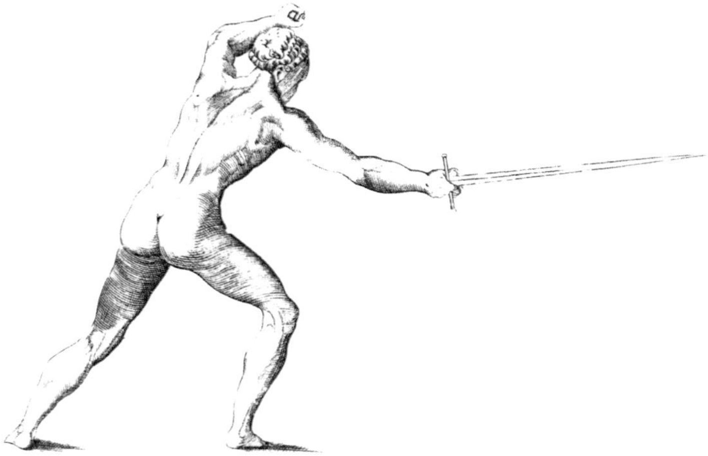

The 4th er

There was a fencing technique called an '*inquartata*', whose name derives from these positions. The fencer performing the technique throws their body out, behind the cover provided by the hilt of their sword. The term '*inquartata*' is derived from '*quarto*', and has a meaning along the lines of 'en-fourth-ed'. To tie it to the Agrippan guard that gives it its name, and to keep it as a piece of fencing jargon, I've translated it as a '*4ther*'.

Translation

fp. §1
RULES OF MANY KNIGHTLY EXERCISES

fp. §2
Collected by Captain Federico Ghislieri, in service of the Most Serene Lord RANUCCIO *Farnese, Prince of Parma, & Piacenza, etc.*

fp. §3

fp. §4
IN PARMA. From Erasmo Viotti's Press. 1587.
With the permission of the Superiors.

p. i §1 [blank]

p. ii §1

p. ii §2

TO THE MOST SERENE LORD RANUCCIO FARNESE, PRINCE OF PARMA, AND PIACENZA, AND SO FORTH.

p. ii §3 *The grace, Most Serene Lord, that Your Highness has deigned to bestow upon me by employing my services on the occasion of practising arms, has reminded me that my efforts with my person alone on that occasion were insufficient. Therefore, it has become my duty to compile a collection, almost in summary form, of all that, both in Theory and in Practice, which I've willingly strived to explain to you, both through words and actions, in obedience to you.*[1]

For I've done this not out of a belief that I perfectly know the profession of using arms
p. iii §1 *(for that isn't truly my profession; mine is that of the military), nor of being able to repay your grace by this means: for in this, I'm certain to fall short in many things, even if I can consider myself very fortunate, and rejoice within myself for being the creation of the Most Illustrious Lord Silvio Piccolomini,*[2] *to whose most courteous humanity towards me in this regard I owe the greatest thanks: in the other, how could I ever repay Your Highness for what I've received, when a thousand more efforts from me wouldn't suffice?*

But I wished to do this so that you might have a sign from me of my most affectionate inclination to continually serve you, and to dedicate myself to you with my efforts, as I'd always do with my life. I add to this a most ardent desire, which I share with your other most faithful servants.

This most ardent desire, at this first bloom of your age, is to assist you in whatever
p. iv §1 *way I can, providing, even through this path I've shown, an occasion for you to become accustomed to the exercise of the body: which, as those who've particularly dealt with this matter have shown, greatly benefits to the securer preservation of life.*[3]

---

1 The original text is a solid block of italic rather than paragraphs.

2 Silvio Piccolomini was a military officer attached to the Medici court, whose career was entwined with Ghislieri's. For more information see the Timeline in the appendices.

3 The Folger copy of the book has a note written in the margin here, which is difficult to read. I believe

This is demonstrated by so many gymnasiums and other places already built by the ancient Emperors (as is read; and even in our times, traces of them are seen) in Rome and in other parts of the world: so that they themselves and their peoples, by exercising in various ways, might gain that vitality which arises from exercise not only for the body but also for the mind: from which benefit it's then easier to acquire the aptitude to employ oneself in both matters of peace and war.

For if we can achieve this gain with a thousand other occasions, we can very well achieve it through the practice of this art of fencing: from which we obtain two fruits at the same time: one is what I mentioned, the vigour and robustness of the mind and body. The other is, whether done in play or in earnest, the ability to defend oneself against the attack of anyone: which is the principal aim of one who practises this art.

Therefore, Your Highness, kindly accept this my effort with the same favour with which
p. v §1 *the most cherished things are usually accepted; for surely there can be no more cherished thing to us than that from which both the preservation of life and the recreation of the mind can be derived.*

Thus, you should accept this my gift, accompanied by the most faithful counsel (for it's sometimes not unbecoming for a servant to give good counsel to his Lord), all the more willingly as we must be certain that you'll do so, following the example of the Most Serene Lord Duke ALESSANDRO*, your Father: who, undoubtedly, in the same age as Your Highness, by dispensing his time in exercises of this most honourable quality, finally reached that greatness of name and deeds of which not only all of Flanders is a witness today but also all of Italy, and almost the whole world marvels at his incomparable valour.*

To which, judging by the evident beginnings now seen in Your Highness, there's no-one
p. vi §1 *who fails to hope to see you walk the same path in parity. May Our Lord God graciously accomplish this in Your Most Serene person; and may I completely witness it. And with the most humble reverence, I kiss your hands.*

p. vi §2 *From the Palace of Your Highness in Parma, the 22nd of April, 1587.*

p. vi §3 Your Most Serene Highness'

p. vi §4 Most humble and devoted Servant,

p. vi §5 Federico Ghislieri.

p. vii §1 [blank]

---

it says '*Di questo mia ci ho da 2 lato tito a pag. 14 pio*', or 'I've a second title from this on page 14.' Perhaps a reference to the Bonelli introduction of the British Library's copy?

p. 1 §1 RULES OF MANY KNIGHTLY EXERCISES

p. 1 §2 *Collected by Captain Federico Ghislieri, in service of the Most Serene Lord* RANUCCIO *Farnese, Prince of Parma, and Piacenza, etc.*

p. 1 §3 I've undertaken to treat various exercises in arms, which, although different
p. 2 §1 from one another, all contribute to the proficiency and nobility of the Knight. It's fitting for the Knight to know how to wield the sword above all other types of arms, both because it's their most familiar weapon, but also because it's specifically designed to defend their honour. Ultimately it's the weapon that aims for victory and brings honour in deeds of arms — whether in sudden assaults or in body-to-body in the barriers.[1] Therefore, we'll begin with the sword as the foundation of all our exercises. To avoid confusion, as many who've dealt with this art have done, we'll establish a method that conforms to human nature, especially when altered by the powers of the soul and its passions, which sometimes prevents a man from acting rationally.

p. 2 §2 The human body is composed of four simple elemental bodies, in which the four elements are combined in their proper forms and natures within.[2] However, when they converge, they possess their respective qualities, which encompass the virtues of these natures. For instance, the cold and dry nature of the earth generates a humour in humans called melancholy, which is also cold and dry. Water, being cold and humid by nature, creates phlegm. Air, being hot and humid, forms blood. Finally, fire, being hot and dry, produces choler.

p. 2 §3 Yet man, according to the common nature of the species, has a body of such
p. 3 §1 temperate complexion that it does not lean towards any extreme. This is unlike animal species, whose complexions are greatly inclined to extremes and are all naturally very subject to certain particular passions, according to the common nature of each of their species. Thus, we see that all hares are timid, all lions are bold, all dogs are irascible. But men, according to the common nature of the whole species, are generally neither timid, nor bold, nor irascible, nor very subordinate to passions.

---

1 *steccati* – an archaic word meaning the space enclosed for tournaments and jousts.

2 This "Hippocratic" or "Galenic" model of four bodily humours was a medical paradigm inherited from antiquity. The first modern book of anatomy, Andreas Vesalius' *De Humani Corporis Fabrica Libri Septem*, was only published in 1543.

p. 3 §2 Nevertheless, even if they lack inclinations according to the common nature of the species, they do have inclinations according to the particular nature of each individual man. Thus, many are naturally irascible because they've a predominantly choleric humour. Others, because they've an abundance of blood, are naturally cheerful and bold. Yet others, because they've an excess of melancholy humour, are naturally sorrowful and timid. Hardly anyone is found in whom the humours are so measured that the complexion is entirely balanced and equal. Thus, it happens that we're all inclined to one passion more than to another, according to the predominant complexion within us.

p. 3 §3 From this arises the fact that if a man is melancholic, their movements will be as though born from the earthy element — pendulous, restricted, anxious, and wearisome — just as the earth itself is pendulous, heavy, and restricted.

p. 3 §4 The movements of water, because they also fall, though not as much as those of
p. 4 §1 earth, are nevertheless descending. They're less restricted and result in phlegm, which corresponds to the element of water. In bodies where it predominates, the movements are timid, simple, and submissive, so the limbs of the body remain almost abandoned and inclined downward. Due to the pallor it infuses in the face, phlegm corresponds to fear or terror, and its bluish whiteness in humans indicates sorrow.

p. 4 §2 The movements of air tend towards the upward, albeit in a moderate and undistorted manner, since it's a pleasant element. The motions of human blood correspond perfectly to these movements: temperate, modest, and regal. The passions of the soul, such as love, which gives rise to delight, pleasure, desire, and hope, correspond harmoniously to these same movements.

p. 4 §3 Lastly, fire naturally has its movements tending, as can be seen from its flames, to the extreme height, twisting as they rise. Similar to these are the movements of choler, for they're violent, impetuous, and fierce. Thus, the passions of hatred and anger perfectly manifest in bodies where this element predominates.

p. 4 §4 As art can surpass nature, we'll strive, through the art of fencing, to make the
p. 5 §1 timid bold, and to maintain the audacity of the courageous. We'll also diminish the anger of the irascible, so that it does not lead to recklessness.

p. 5 §2 And although fear, by its nature, greatly corrupts judgement and impedes counsel; and although anger has the same effect, it will still be easy to make the timid man bold. This can be achieved by reducing fear, which, when it's not overpowering, brings with it carefulness. For if the timid man, through reasoning about things, becomes capable at arms, they'll become more courageous. For the source of fear in a coward is their lack of efficacy.[1]

1 The original here is '*virtu*', but used in the archaic sense of vigour, strength, or power of action.

p. 5 §3 Similarly, we'll moderate the anger of the irascible person and make them more apt to use reason. By offering good counsel, we won't let them be carried away by passions. We'll show them how to carefully weigh everything that can help or hinder them, and how they should approach dangers with cautious steps and enter them slowly.

p. 5 §4 It's also worth noting that through the imagination of those things that make our endeavours seem easy, great hope arises in us. From this, we become bold due to the advantages that are absolutely within our power, such as strength and the art of fencing. Hence, we act courageously and, with reason, achieve our aim, which is victory alone.

p. 6 §1 *Therefore, to show this ease in our present art, we'll begin to discuss its Theory: After which we'll discuss the Practice, most necessary to this profession.*

p. 6 §2 Firstly, we must know that seven circumstances, or particular conditions, are required for human actions: for although they aren't intrinsic parts of human actions, nevertheless they're always necessarily around the actions of man, in such a way that he can't perform any without them being present.

p. 6 §3 It will therefore be necessary for the understanding of this our science of arms that we consider the aforementioned circumstances:[1]

	Circumstance of Action	*Circumstance in the Science of Arms*
	1. The agent.	Man.
	2. The work, or action performed.	The act of stabbing.[2]
	3. The matter, around which we act.	Movement.
	4. The instrument, with which we act.	The sword for offence, and the dagger for defence.
	5. The location, where it takes place.	In public streets.
	6. The manner in which we act.	The manner in which we attack others, and defend ourselves.
p. 7 §1	7. The purpose for which we act.	The goal of victory.

p. 7 §2 The human body is found to be composed of such measured proportions, with each part so well corresponding to the whole, that the ancient architects derived the composition of almost all things from this same proportion, such as building houses, churches, castles, ships, and in short, every type of structure. This is written by all the ancient and modern authors who've dealt with this subject, among whom Vitruvius, in particular, stands out at the start of his Book 3.

p. 7 §3 This proportion of man was judged optimal for this purpose, as I say, by these capable men, although there's no certain and determined proportion in the stature of the man, as some men are taller, and others shorter. If a man is to be of a suitable stature, they'd have to be 6 human feet tall, for Vitruvius himself says that the foot is ⅙th of a man's stature.

---

1 The table is an editorial change. The original has sentences listing the causes in the preceding two paragraphs. The concept of human action having accidental circumstances was well embedded in Western philosophy, see Aristotle's *Ethics*, Book 3, Chapter 1, Cicero's *De Inventitione*, Book 2, Chapter 27, or Aquinas' Summa *Theologica*, Book I-II, Question 7, Article 3.

2 *coltellate* – attack, thrust, stabs; literally 'knifings'.

p. 7 §4 In this regard, Vegetius, in the first book of his art of war, states that the
p. 8 §1 consul Marius selected recruits, that is, new soldiers, who were 6' tall, or at least 5'10".[1] Hence, it follows that a man shorter than 6' was considered of average stature, and 1' taller than this was considered very tall. Furthermore, when a man exceeds the height of 7', he's deemed supernatural and called a giant, according to the rule of Marcus Varro, as Aulus Gellius recounts in the Book 3 of *Attic Nights*.

p. 8 §2 This is in accordance with what Suetonius says in the *Life of Octavian*, speaking of the stature of that Emperor. He was of medium height, being 5'9".[2] But this average height wasn't noticeable unless he was next to someone tall.

p. 8 §3 The ancient philosophers, upon speculating, have discovered that the shape of a circle, which is the most perfect of all others, is found perfectly in man. This fact is can be understood like this: let a man lie down with their face up and stretch out their arms and hands, as well as their legs and feet, as much as they can. Take a compass or the measure of a foot of the same man, and measure it from the navel as the centre: then a round and perfect circle will result. This has been noted by Vitruvius in the aforementioned Book 3, and he demonstrates it with the following figure:

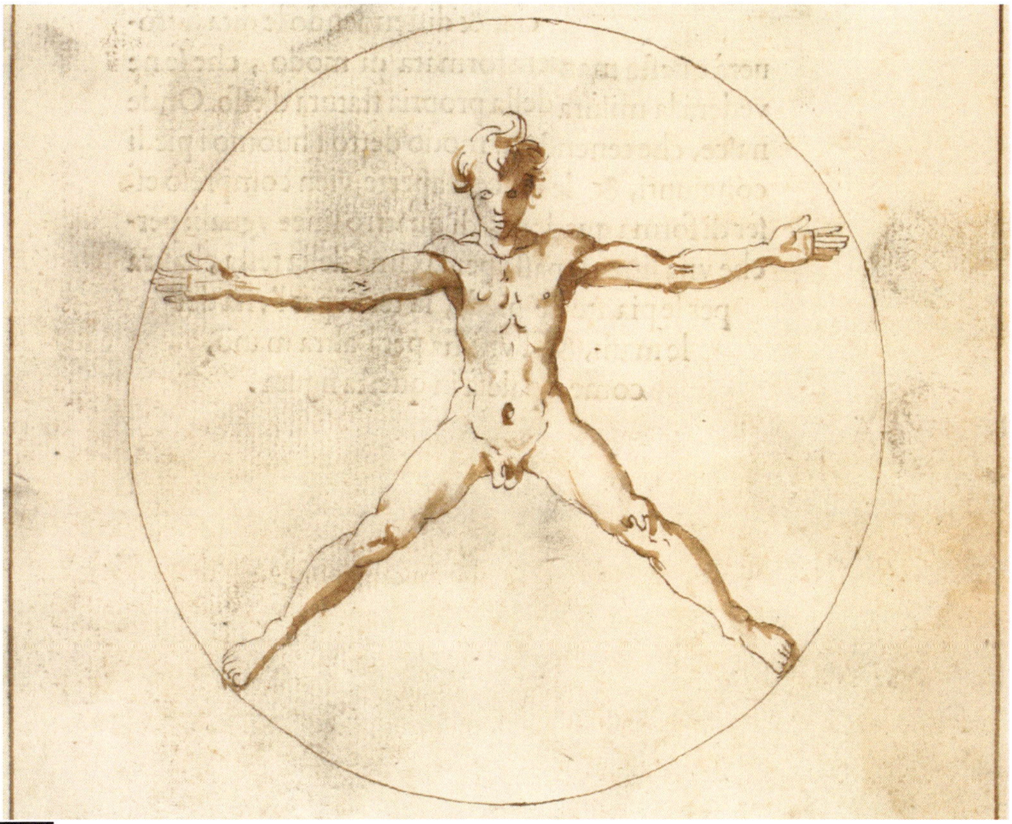

1 The original explains that 10 inches is $^{10}/_{12}$th of a foot.

2 Again, the original explains that 5¾ foot is the same as 5'9".

p. 10 §1 Pliny, likewise, in Book 7, Chapter 17 of *Natural History,* writes that humans are formed in a square, angled, shape. That shape is made this way; by extending the arms and spreading the fingers, the measure of the person's own height is shown. Hence it follows that, by holding the feet together and the arms extended as described, the man is understood to be of a square form, with four equal sides: because one line passes over the top of the head, another over the soles of the feet, the third over one hand, and the last over the other hand, as illustrated in this figure:

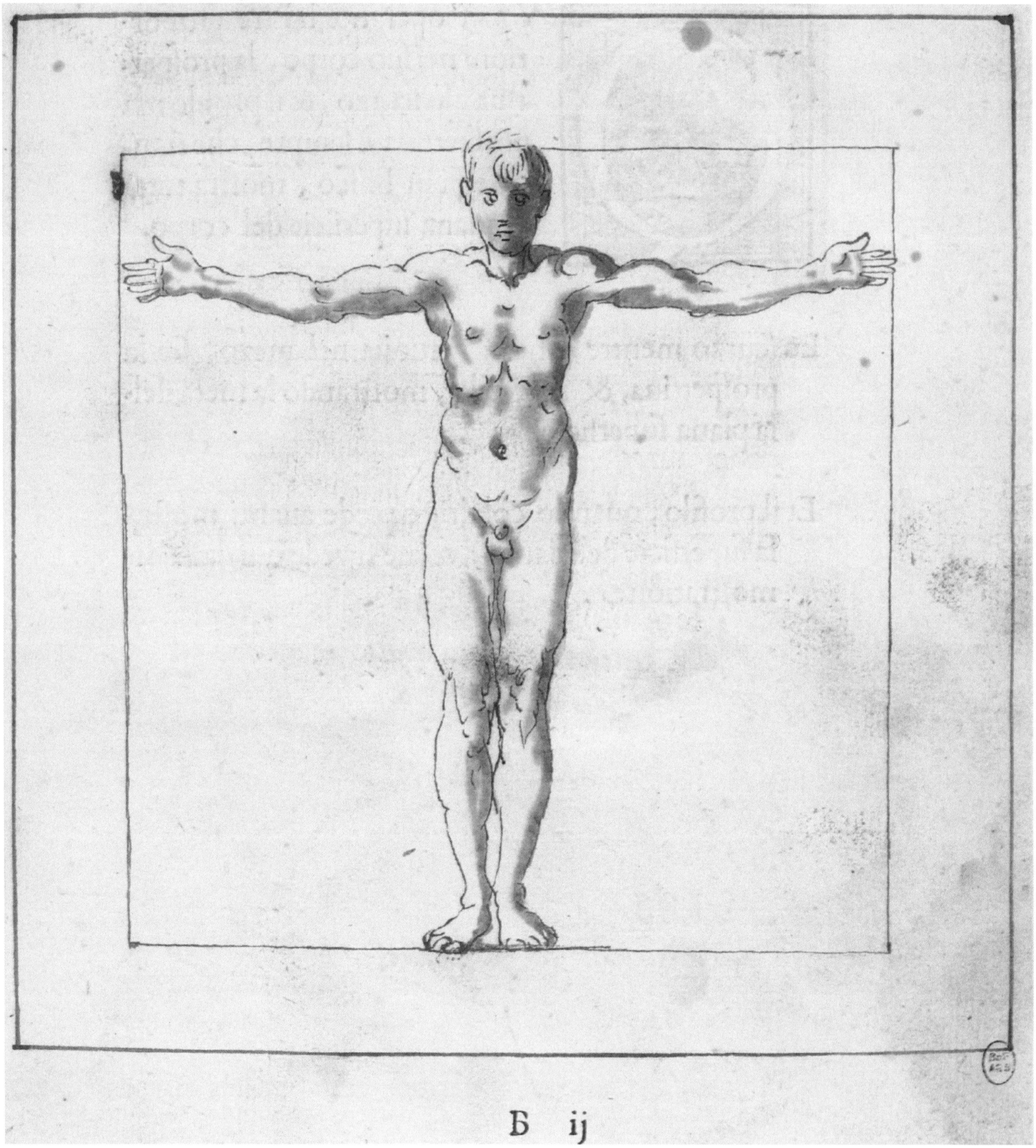

p. 12 §1 Our actor's body has three proportions: facing, foreshortened, and in profile.

- Facing is whenever, standing in balance, he shows the full flat surface of the body.

p. 12 §2
- Foreshortened is when it's found in the middle, between facing and profiled, showing half of the flat surface.

p. 12 §3
- Then profiled is when, with the right foot forward, they show the side's surface, as seen here in the illustration.

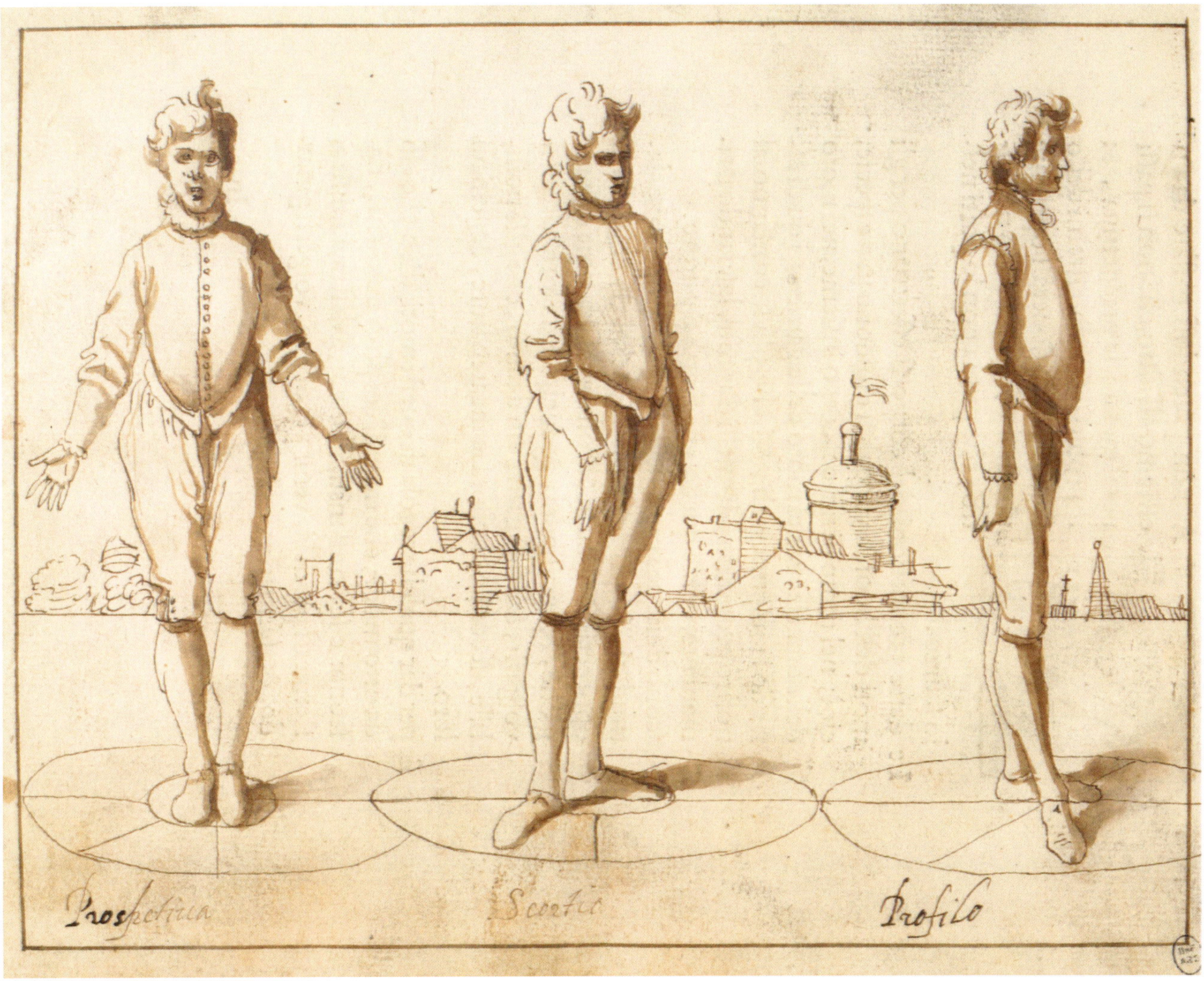

Facing *Foreshortened* *Profiled*

p. 14 §1 But since it's impossible to move from one extreme to the other without passing through the middle, and since facing and profile are the extremes of foreshortening, the operative can't move the body from one to the other without first passing through foreshortening.

p. 14 §2 The aforementioned proportions of the body are realised through the movements of the man, which are divided into three types:

1. The movement of the entire body together.
2. The movement of the arms.
3. The movement of the legs.

Since these are the most important and most relevant to our purpose, we'll discuss them and ignore the rest, since they're the domain of the painter.

p. 14 §3 Thus, the movements of the body arise from the lengths, widths, and proportions of the limbs; and from their pressing, rotating, and corresponding together with reason and possibility; and also from their twisting, turning, and extending as much as possible, according to their joints and pivots.

p. 14 §4 Rationalising such movements, I say that they arise from the eight ways that the body can move: upward, downward, to the right, to the left, extending outward, retracting inward, turning by rotating, and stopping.

p. 15 §1 In every movement, whether forward or backward, it's necessary for the body to be supported by its legs; otherwise, due to the nature of heavy bodies to tend towards the centre, it'd fall.

p. 15 §2 Thus, it can be derived that the movements of the head are such that a man can barely turn it in any direction, without it always being supported by some part of the front of the body positioned beneath such a heavy weight; or else, like a set of scales, he places a limb on the opposite side to balance the weight.

p. 15 §3 For the same is observed when someone, having stretched out the hand, supports a weight: because having fixed the foot as the foundation of the balance, the entirety of the rest of the body counterbalances to equalise the weight.

p. 15 §4 When a man stands with feet together, and makes an arch forward with the whole body and legs, the tip of the chin will remain straight over the tip of the feet.

p. 15 §5 When pressing, either with the right hand or with the left hand, the crown of the head will always be straight above the said feet.

p. 15 §6 Whenever the body leans on the side of the foot that's weight bearing, the shoulder will be straight above the ankle, and the other leg will counterbalance the body.

p. 15 §7 Whenever one raises an arm high, all the other parts of the body on that side
p. 16 §1 down to the foot follow that upward motion, so much so that even the heel of that foot will lift from the ground due to the movement of the same arm.

p. 16 §2 A limb never extends in one direction without the others following it; nor does it press or contract without the others following almost like lines towards the centre.

p. 16 §3 The legs, or feet, perform three types of motion: straight, circular, and sideways.

p. 16 §4 1. Straight motion occurs whenever the body is facing, and the legs open directly, as when a step is made.

p. 16 §5 2. Circular motion occurs whenever, with the body standing still on one foot, describes a circumference with the other.

p. 16 §6 3. Sideways motion occurs whenever one stands facing, and moves either to the right or to the left.

p. 16 §7 The legs in profile can't open as much as the length of the body.

p. 16 §8 This line or extent described by the opening of the body's compass could be divided infinitely: but for the sufficiency of our art, we'll divide it into four parts: half a step, one step, one and a half steps, and two steps.

p. 16 §9 When one leg is placed, the other won't be able to move forward further than the space of one foot.

p. 16 §10 From these compass openings, it follows that the body can be found in a
p. 17 §1 stance or at rest in five ways:

p. 17 §2 1. When it's balanced. That is with the feet together, for at that moment the weight is directly above is in the diameter of its circumference. For in such a stance it's ready to move in all directions.

p. 17 §3 2. When a man stands with the whole weight of the body on one foot, in the manner of the base of a column, meaning the ankle, is placed straight under the hollow of the throat. This posture was invented by the ancient Polykleitos.[1]

p. 17 §4 3. When a man is at rest in a step.

p. 17 §5 4. When the body is in one and a half steps, forming an equilateral triangle.

p. 17 §6 5. When the body stands in two steps, known as a forced step.

---

1 Polykleitos, or Polycletus, was a C4th BCE Greek sculptor, credited with inventing the *contrapposto* position. *Contrapposto,* or *counterpoise* depicts a person with their weight on one engaged leg, while the other is free and bent. This is in contrast to the earlier, more rigid forms of archaic sculpture.

p. 17 §7 From these rest positions, a man forms six circles – two without moving, and four after describing one of the mentioned quantities and forming a circle:

p. 17 §8 1. The first of the two [stationary circles] is formed when, standing balanced, a circle is described around their feet, the centre of which has weight straight above.

p. 17 §9 2. The second [stationary circle] is when, standing in said [first] circle, a
p. 18 §1 [second] circle is described at a distance equal to their height from their body; the centre of which will also be the weight straight above, as seen in this figure:

p. 19 §1 1. The first of the four [circles of motion] is formed when a man is in the second stance of rest (as discussed); when, with the stable left leg (on which the weight of the body sits) he makes a centre with it and describes a circumference with the other mobile leg. We use this to keep ourselves in guard.

p. 19 §2 2. The second [circle of motion] is when they're in a step. Standing still, they keep the left foot stationary and describe another circumference with the mobile other [foot]. In this way we do sideways movements, remove our points.

3. The third [circle] is similarly formed, while in the fourth stance [with one-and-a-half steps, forming an equilateral triangle], which we enter when making the strike.[1]

4. The fourth [circle] is when he's in the fifth stance of the forced step: as shown in this figure: and with it, we keep the enemy at a distance.

1 Derived from p. 17 §5.

p. 21 §1 *Having discussed the agent or artisan, it's appropriate to proceed with describing the material, since the action of striking*[1] *is evident in itself.*

p. 21 §2 Since it's movement, and matter, by which man operates in this action of fighting with blades, we must understand that movement, as Aristotle defines it in his *Physics* (Chapter 9, Book 5), is a change or transformation. Some believe there to be six types of change:

1. Generation
2. Corruption
3. Augmentation
4. Diminution
5. Alteration
6. Movement, from place to place.

Nevertheless, Aristotle himself concludes in the aforementioned text that there are three and no more:

1. Change of Quantity
2. Change of Quality
3. Change of Location.

Of these three types, the last is the one we need to understand for our art. This movement is nothing but the transmutation that occasions a body moving from one place to another, and the end of movement are two instances.

p. 21 §3 The instant in movement and the instant in time are like the size of a geometric
p. 22 §1 point, that is they've no parts but rather are indivisible. Consequently, instants are neither movement nor time. However, they're indeed the beginning and the end of every movement and every finite timing, as Aristotle explains in *Physics* (Chapter 24, Book 6).

p. 22 §2 Movement sits between two states of rest - and rest, or repose, is nothing but the absence of movement.

p. 22 §3 And there are three types of movement: two simple ones, which are straight and circular, and the third one, which is composed of these two, making a composite line.

p. 22 §4 The two simple movements are either natural or forced.[2]

p. 22 §5
- Natural movement is what heavy bodies do from a higher place to a lower one, straight down, and without any effort.

1 *coltellate* – attacks, stab, or knifings. Florio's 1611 dictionary gives *coltellare* as 'to fight or fence with knives, or swords or any cutting weapon', and *coltellata* as 'a blow with a knife or any cutting weapon'.

2 *violento* – requiring strength.

p. 22 §6
- Forced movement is what bodies do with effort, from below to above, or from side to side, due to some moving power.

p. 22 §7 Natural movements are weak at their beginning, and the more they continue, the stronger they become.

p. 22 §8 Forceful movements are very powerful at their start, but the farther they get from the beginning, the weaker they become, losing their strength.

p. 22 §9 There are four types of movement from place to place:

1. Pushing, whereby we move things away from us, pushing them to another place, as we do. when pushing the sword forward in a thrust.
2. p. 23 §1 Pulling, where (contrary to pushing) by pulling and moving the object toward us, we draw it closer; such as when we pull back the sword towards ourself.
3. p. 23 §2 The third movement is called carrying, where we neither push away from us nor pull towards us, but move by carrying along with us, such as when we carry the sword fixed in a position with the motion of our body, moving it now here, now there.
4. p. 23 §3 The fourth movement is called revolving or rotating, where, by moving something in a circle, bringing part of it toward us and part away, we turn it in such a manner that this movement is almost a combination of pulling and pushing, as seen in circular strikes.

p. 23 §4 As a man's body has five centres of movement, so there are five circumferences that form with striking movements:

1. The first centre is in the left foot. By standing it still, raising the right arm and by moving the entire body, it describes the largest circumference that they can make. This forms the cut.[1]
2. p. 23 §5 The second circumference is formed by the man when, having made the body's waist the centre of a circle, he describes a circumference with the rest of the body and the arm raised, which is used by many in fighting at the barrier.
3. p. 24 §1 The third circumference is made by the man when they find themself in one of the five aforementioned stances. While keeping their entire body still and moving only the arm they describe a circumference with the shoulder as the centre. This is used for executing direct strikes while in a stable stance with the legs.

1 That is, making a cut with a passing step; starting with the left foot in front to pivot around its position.

p. 24 §2 4. The fourth circumference is described by the man when — with their entire body still and using the elbow as the centre — they form a circumference by moving the remaining part of the arm, called the forearm. This is used for avoiding hits when the enemy tries to parry.

p. 24 §3 5. The fifth and final circumference is made by the man when — with their whole body still and using the wrist as the centre — they move only their hand to describe the smallest circumference they can. This is used for creating knots of the hand.

p. 24 §4 *Following our sequence it remains for us to discuss the offensive and defensive instruments.*

p. 24 §5 The offensive instrument is the sword, and the defensive one is the dagger. Sometimes the sword may serve the function of the dagger by parrying, and the dagger may serve the function of the sword by attacking; but this happens by accident.

p. 24 §6 The offensive instrument, the sword, is divided into an honest[1] edge and false
p. 25 §1 edge, both converging to a point. Although it's a body contained by a surface that could be divided in all three dimensions, we must consider it in its operation as a line.

p. 25 §2 There are four ways to hold the sword:

1. Firstly, by placing the thumb on the spine so that it helps to cut more accurately. However in this manner the point–work is poor, and the hand does not exert its full strength, being half–open.

p. 25 §3

2. In the second way, one might hold it flat with the palm of the hand facing the ground. This way, it feels lighter in the hand, since the hand and arm are in such a state almost acting as a support for the body, which relieves the weight of the sword. But in this manner, you can't strike with the edge. Rather, it's played by changing lines [feinting] which is quite bad.[2]

p. 25 §4

3. In the third way, the sword is gripped with a clenched fist, in order to gain a finger's length. Indeed, this is true with the effect to the instrument, but every time a strike is made the instrument should form a straight line with the arm as much as possible, which is unachievable with this grip. It always forms an obtuse angle. Besides the fact that the cuts mostly strike with the flat of the blade.[3]

p. 25 §5

4. In the fourth and last way (which is the best one), the sword is held with
p. 26 §1 a clenched fist, and the index finger across the cross of the hilt. This way of gripping is best.

p. 26 §2 The sword, when hitting, can be likened to a wedge. The heavier the weight that strikes, the greater the impact will be. Furthermore, the further the distance between the striking weight and the wedge, the greater the impact will be.

---

1 The word here is '*buono*', also meaning good.

2 This method is one advocated for by Swetnam in *The Schoole of the Noble and Worthy Science of Defence*, 1617.

3 I understand this to be grasping the bottom of the hilt for range, like pommelling a French-grip epee. For a more in depth interpretation of this section, see https://blog.subcaelo.net/ensis/gripping-the-rapier/

p. 26 §3 Thus, the heaviness of the sword [impact] can be increased beyond its own weight, by a driving force.

p. 26 §4 By itself, whenever the sword is moved from a distance, its weight increases in proportion to the speed. This is because every heavy object, while moving, becomes heavier than it'd be stationary, especially when moved from a greater distance.

p. 26 §5 As for the driving force, as we've said the man can form five circles, through which the weight increases. Having formed the first, which is the largest, the others follow and the weight always decreases because the circumferences moved will be smaller [with each circle].

p. 26 §6 The defensive instrument, which I take to be the the dagger, is a resistant
p. 27 §1 body, whose job is to defend the body of man. It can be held in the hand in three ways:

1. In the first, it's held flat in the hand, with the thumb in the middle of the blade; and by dividing the body into two parts with a semicircle of natural movement, it defends the right part; and with another semicircle, it defends the left part: and therefore it must be put in position, so that with these movements it makes the said defences. This has many flaws: one is that the hand is subject to attack. Another is that, although a small circumference is described by moving the last centre, the line of the dagger forms an angle with that of the arm, through which many lines can enter. Furthermore, it's subject to many inconveniences both inside, as outside; and so below, as above.

p. 27 §2

2. In the second, the dagger is held the same way, but brought closer to the right side. It only seeks to beat outwards, so as to prevent a strike before it reaches the body. This is poor because instead of covering, it exposes. It's also a forceful motion. Moreover, every time the fist forms a little angle, a dagger held like this won't prevent that sword from entering. It's also subject to deception.

p. 27 §3

3. In the third and last (which is best), the dagger is held in hand in such a
p. 28 §1 way that with it it can strike when necessary. From there, with natural movements, it defends the body by covering. Held thus, always free from being attacked and it's not subject to deception.

p. 28 §2 Moreover, this defending line will always dominate the opposite sword from the outside; and with natural movements will beat, always helping their line to descend. For even very little force adds new motion to any weight if it was already moving.

p. 28 §3 However, the last third of their sword will be beaten, for it's the weakest part in this effect. That said, it's the strongest when striking, for the first part of any inanimate thing is easier to move in the other part, when moving with a forceful motion. This should be understood in this way, since everything moves in a continuum it's much easier to move something aside from the end opposite from that which is being worked upon. That's because the other end is moving at great speed.

p. 28 §4 Yet, as with other things that are thrown or pulled, the motion weakens in the end. Thus at the end of the continuum, the motion becomes weaker and there's less resistance. Things are more easily pushed from that end in which there's greater weakness (and therefore less resistance), which is undoubtedly the purpose.

p. 28 §5 This defensive dagger must form a straight line with the arm, and it must
p. 29 §1 also form a right angle with the body's surface. In this position, the man's view should pass through the strong point of the dagger and fall on the weak of the enemy's sword.

p. 29 §2 Acting in this way with the dagger, we'll necessarily find the diameter of the circle of the enemy's body with our sword.

p. 29 §3 Since the dagger must always dominate the enemy's sword, it must be noted that when our enemy raises the sword, we'll raise the dagger in such a way that our sight, passing through it, sees their sword dominated inwards.

p. 29 §4 The dagger beats at straight lines, taking them aside from their path, and beats all angles outwards. Thus it beats the opponent's line in three ways:

1. Firstly, voluntarily.
2. Secondly, out of necessity.
3. And thirdly, deeply.

In the voluntary one, it's beaten inside, pulling their blow out to a greater length.

p. 29 §5 In the necessary one, it's not beaten but only by resisting the strike do you escape with the body, getting out of the strike by turning the false edge towards the ground.

p. 29 §6 For the deep one, it's beaten with a semicircle outwards, when the opponent's
p. 30 §1 sword is parried in a sideways line in presence, under the centre of the extended body, having to pull the furthest length to strike. If the end of the line made the opposite motion, the said length couldn't be obtained.

p. 30 §2 It's not necessary with the defending line to bring the opposite line to the centre from where it originates, but one must beat it when it's found there. Yet when it's not there, one must give it help, so that the natural motion reaches its lowest more quickly.

p. 30 §3 *Now that we've discussed the offensive and defensive instruments, we'll proceed to talk about the location.*

p. 30 §4 The location will be any kind of place. Since most often fights and duels occur in the public streets of cities, which are often hazardous due to rain, mud, and being uneven, therefore we must adapt to every type of site. We'll achieve this by keeping the body together and well-distributed in its bases.

p. 31 §1 The sixth condition, which is the manner in which we offend others and defend ourselves, primarily consists in securing our body from the enemy's attack. Therefore, the attack will never be made unless we're first sure of our defence; nor will we ever defend ourselves simply without, at the same time, attacking. This is because the true defence is to offend, which will be done resolutely; and we must always be the first to execute the strike.

p. 31 §2 From these offences, two kinds of strikes arise, which are cuts and thrusts.

p. 31 §3 The cut has all these properties:

- It's more natural to man than the thrust, given that man's movements are circular.
- It targets a larger portion of the opponent's body.
- It's a natural movement.
- Being more visible, it induces fear.
- It can also end at any point you want.

p. 31 §4 Nevertheless, it has these flaws, namely:

- It exposes the right side of the body and the same arm
- It's not lethal, since it's resisted by bones and armour.
- Additionally, due to the shape of the sword, which is not spherical and therefore not equally heavy, it often strikes flat when resisted by the air, and the sword is subject to breaking.

p. 32 §1 The thrust is delivered with the body covered and lands before it's seen: therefore, it's more damaging and involves less movement. Due to the shape of the sword, which is wedge-shaped, it strikes more securely and is more lethal; hence it's better to use than the cut.

p. 32 §2 The cut, then, is divided into three simple types and two compound types.[1]

1 I believe this section draws on dall'Agocchie (1572).

Of these three types of cut:

1. the first is cleaving.[1]
2. the second is across.[2]
3. the third is around.[3]

Each of these three types is further divided into forehand and backhand:[4]

- Forehands are those that come from the right side
- While backhands are those that come from the left side.

These forehands and backhands divide the circle of a person into eight equal parts, as shown in the following figure:[5]

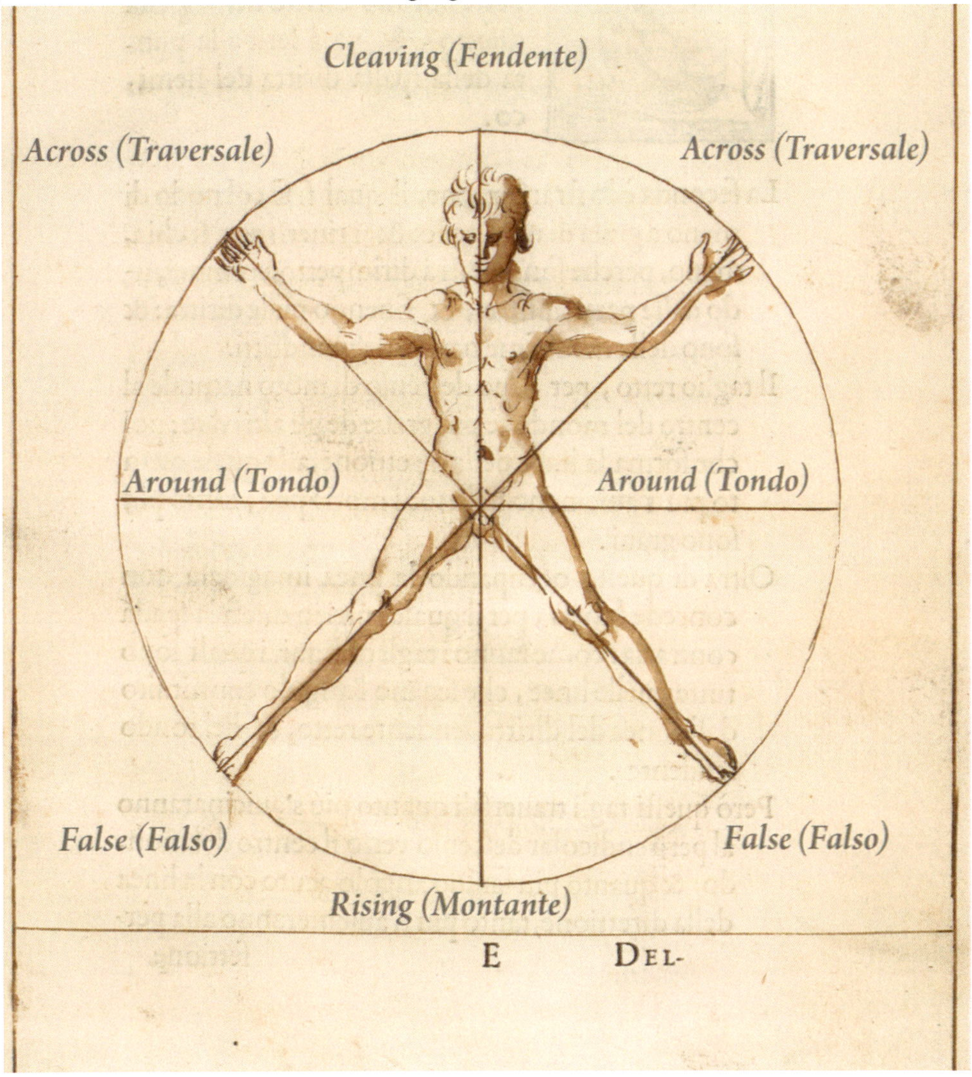

1 *fendente* – cleaving.

2 *traversale* – across.

3 *tondo* – around.

4 *diritti* is literally 'straights' or 'rights', and *roversi* is 'reverses', but in every day parlance, forehand and backhand.

5 The Bologna copy is labelled with names for the cuts (reproduced here).

p. 32[1] §1 Of the two compound types:

1. One is the redoubled forehand cut, which starts with the good edge of the sword below, and strikes at the tip of the enemy's right shoulder.

p. 32 §2 2. The second is the overwhelmer,[2] which is performed with the wrist in a circular motion.

The backhands are called such because they're opposite, starting from the left side and ending on the right: they're of the same nature as the forehand cuts.

p. 32 §3 The straight cut, due to its natural descending motion toward the centre of the world, is heavier than the other two – because it forms the straight line down. The closer that weights approach this line in their motions, the heavier they become.

p. 32 §4 Furthermore, it occupies space. The aforementioned line does concede space for the opposing sword to enter in defence, as the sideways cuts do (which are all those lines that bisect the angle formed by the straight downward cleave and the cleave around.)

p. 32 §5 However, for these cuts across, the closer they come to a straight downwards
p. 33 §1 path, toward the centre of the world, and the more they form an acute angle with that line of direction, the closer they'll come to perfection. Conversely, the more they deviate from it and approach the horizontal line of the right circular cut, the less they'll form an acute angle; consequently, they'll be more slanted and, therefore, of lesser weight and force.

p. 33 §2 The circular downward cut can be sufficiently understood from the description of these two types: and everything that's been said about the forehand cuts should also be understood for the backhand cuts, since they're of the same nature.

p. 33 §3 All these cuts can be executed by striking with the wrist in a continuous motion. That is, aided by the moving force from the beginning to the end. The man relies more on the rapid strength of their body, rather than on the force of the motion together with the body. This is further assisted by the chopping action, which makes the wound bigger. This approach has some apparent advantages: besides, the person striking in this manner is more protected, deviates less, and performs the action more precisely.

1 Here the printers made an error, repeating page numbers 32 & 33.

2 *Stramazzone* – literally 'over-macing', used by other fencing authors such as Vadi (1480s) and Capoferro (1610), who defines it as cut made with a turn of the wrist on p. 40.

p. 33 §4 But since the strike is an extremely powerful force, as Aristotle states in *Mechanical Problems*, № 19,[1] we shall conclude that all wounds, in order to be perfect, must be assisted by both the strength of the body and the force of the strike.

p. 33 §5 The second manner of striking [thrusting] is divided into three types: namely, the broach,[2] the stab,[3] and backhand thrust.

p. 34 §1 The broach is made overhand, which comes from high to low in a natural motion. When that doesn't stop at its low point, but ends rather at the plane of the shoulder, and forms an equilateral triangle in striking and in recoiling, it's called the "thrown broach".

p. 34 §2 The stab is made underhand with a forceful motion, pushing the sword straight forward. There can also be differentiated the "tapered stab", which is so called because it has the same effect in striking and recoiling as the "thrown broach".

p. 34 §3 The backhand thrust departs from the left side.

---

1 Problem № 19, from https://digitalcommons.unl.edu/classicsfacpub/68/ :

"Why, if you put a large axe on wood, and a large burden on that, doesn't it pull apart the wood, no matter how considerable the burden is? But if you raise the axe and hit with it, you split the wood even if you have less weight than you put on the axe in the first place?

Is it because everything works through moving-power? And the weight receives the moving-power of the weight more upon being moved than while at rest? So lying [on the wood] it doesn't get moved the moving-power of the weight, but receives it, and that of the striker, upon being carried.

Also the axe is essentially a wedge, and a wedge, though small, pulls apart large objects because of being two levers set in opposition."

2 *Imbroccata* – composed of '*im-*' + '*broccare*' + '*-ata*', literally "en-thorn-ed". I've given it as *broach*, the archaic word choice for piercing highlights that this is a piece of fencing jargon. You can always read the original.

3 *Stoccata* – composed of '*stoccare*' + '*-ata*', literally "having thrust-ed".

p. 34 §4 *Because in any kind of wound the sword enters in some proportion to a line, it will be necessary for us to state some geometric principles.*

p. 34 §5 The point should be understood as that which, due to its smallness, is indivisible in every direction: it's not a quantity, but that which is quantified: and it can't be perceived by any external sense, but only by our imagination. The following is a point:

p. 35 §1 The line is a quantity with length, without width, and without thickness or depth. Consequently, it can only be divided lengthwise, being indivisible in every other direction.

p. 35 §2 So the two ends, or extremities of the line, are two points; as can be seen in this example: 

p. 35 §3 There are three types of line; straight, angular, and mixed.

p. 35 §4 1. A straight line is one that extends from one point to another as short as possible. It's as follows:

p. 35 §5 2. An angular, or twisted, line is one that extends from one point to another in a twisted manner: it's as follows: 

p. 35 §6 3. A mixed line is one composed of these two types:

p. 35 §7 Now, to our purpose, the straight line is the one that begins from the left foot with the body in profile, and the arm extended; and when the tip of the sword is aligned with the left shoulder. The sword will also be in a straight line whenever it's in position, ready to strike directly.

p. 35 §8 The sword will be in an angled line when it's positioned diagonally to the opponent's body; that is, when the body of the sword is visible, but not the tip.

p. 35 §9 Parallel or equidistant lines are those that come no closer together on one side than on the other, even if we could imagine them extending into infinity.[1]

p. 38[2] §1 From this, I derive that the strike will make a right angle to the opponent's body: thus the enemy's sword and our body will form a parallel line, and we should similarly aim for the largest possible target on the opponent's body.

1 Here the Folger has two parallel lines, unlike the other versions I have seen.

2 Here the printers corrected their previous error, skipping two page numbers from 35 to 38.

p. 38 §2 A right angle is formed by two lines that touch and are perpendicular to each other.[1]

p. 38 §3 An acute angle is one that's smaller than a right angle.

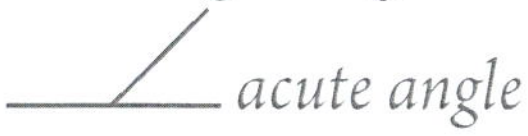

p. 38 §4 An obtuse angle is one that's larger than a right angle.

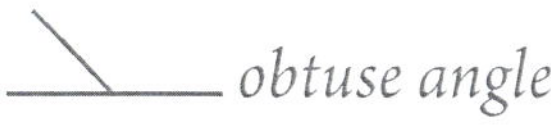

p. 38 §5 These lines extend according to the assistance provided by the whole body and its parts. For example, the more the body leans to form an acute angle at its centre with the line of the thigh and with that of the body, and similarly the more the legs open like a compass, and the more it ensures that the arm with the sword forms a straight line, the longer the line of the sword will become. Thus, because these divisions could be multiplied infinitely, for the purposes of our art, it will now suffice to divide it with the four compass openings already mentioned, as seen in this figure:

1 Only the Folger copy of the book has these diagrams drawn in the margin.

p. 40 §1 They can, however, be extended in four ways:

1. The first, most realistic way is by positioning the body in profile and extending the arm to form a straight line with the left shoulder; nonetheless, such a line does not include defence.[1]

p. 40 §2

2. The second way would extend further if, after positioning in profile, the body's weight were shifted to the right leg, and if the left shoulder were elevated as a counterbalance to the body. However, this would weaken the strike, result in excessive movement, and necessitate a forced return to the other leg.

p. 40 §3

3. The third way involves placing the left foot behind the right one sideways to extend the line a little more. But this is a forceful, uncomfortable, and dangerous move. It risks making us fall and is not secure for retreating.[2]

p. 40 §4

4. The fourth and most perfect way is achieved with the body facing, by bending at the belt, with the compass opened; forming a straight line while turning the false edge of the sword towards the ground. This way extends the line more than in profile and includes defence.

p. 40 §5 To conclude the seven circumstances that pertain to all human actions, it
p. 41 §1 remains to briefly understand the end for which man moves in this action; and this is victory, which, being self–evident, requires no further explanation.

1 Here the Folger copy has a little sketch figure in the margin.

2 This is describing Agrippa's attack in 4th.

p. 41 §2 *As all our discursive understanding stems from suppositional knowledge (from which we can start our discourse), and since there is a need for some further geometrical principles in the rest of our work, we will present those now.*

p. 41 §3 Therefore the Circle (to start with), or circular shape, is a flat shape held by a single line, in the middle of which a point can be taken. All lines from that point, extending to the circle, or circumference, would all be equal. That point is called the centre, and the line which contains the shape is the circumference.

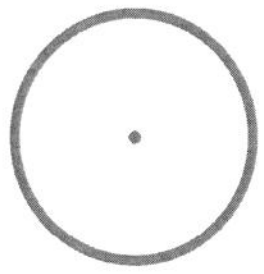

p. 41 §4 The Diameter of the circle is a straight line passing through its centre and touching the circumference at both ends, dividing the circle into two equal parts, or rather into two semicircles.

p. 42 §1 The Semicircle is a flat curved shape contained by the diameter of the circle and half of its circumference.

p. 42 §2 A Chord in a circle is the straight line that divides the circle into two unequal parts, and consequently does not pass through the centre, as seen here.

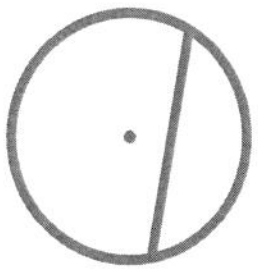

p. 42 §3 Therefore, the larger remaining part is called the major portion, and the centre stays inside it.

p. 42 §4 The part that's left, without the centre of the circle, is called the minor portion.

p. 42 §5 We must know that not only bodies but also surfaces, lines, and points do not have an existence except in some natural matter.

p. 42 §6 It is quite possible that they may be understood by themselves with the imagination, without application to any sensible matter. So for example in a sword, in which not only the body but also the surface, the line, and the point actually exist. Through the imagination the body can well be understood to be contained by a surface, and on that surface, to imagine the lines and the points considered in themselves, and without regard to iron or other natural matter.

p. 42 §7 So in this way we will consider the line of our action as an abstraction,
p. 43 §1 independent of the matter, when the sword is not present; but when it is set in a straight line, then we will consider the line to be applied to the matter of the sword.

p. 43 §2 *Now, moving on to other principles, I tell you that just as it is necessary in other human operations to have seven circumstances, so too in this particular action of arms the same seven circumstances are present. Namely: will, knowledge, measure, timing, occasion, place, and weight.*

p. 43 §3 Will is that which all our actions depend upon: considering that all powers
p. 44 §1 of action, and of all the organs, are like instruments of the will, which is the principal agent. It is in man like a King, who has a principal Counsellor, according to whose advice he knows he must do everything, and this is the intellect. He also has certain other subjects, who are like observers, though sometimes they turn out to be deceivers: and these are all the external and internal senses. Besides these, he has two other subjects, like his lieutenants, who must be ready to await the King's commands, to obey him: these are the desiring and the irascible faculties. Those are powers of appetite: their office is to command movement to the limbs. Lastly, this King has a minister whose role is to execute everything that is imposed by him or by his lieutenants: and this executor is the motive power, which uses the body and its parts as instruments according to its need.

p. 44 §2 Now, to apply this to our subject, I say that it is necessary that in the practice of arms, we have the will ready, with all its officers, so that the motive power operates according to what needs to be done: and it should be more focused on the strike than on the parry.

p. 44 §3 Knowledge is the understanding of something through its causes, for a cause means that from which comes an effect, of which it is the cause, and with which it is possible to adequately assign a reason for it's origin.

p. 44 §4 And just as every question about any effect can be resolved in only four ways
p. 45 §1 by assigning either its matter, its form, the producer who makes it, or the end that motivates its making so there are only four types of causes for their effects:[1]

1. the *material* cause
2. the *formal* cause
3. the *efficient* cause
4. the *final* cause

p. 45 §2 The *material* cause is the subject that remains under the form and does not dissolve from it as long as the form remains intact: for instance, in these matters of fencing, the motion can be considered as an example.

1 These are Aristotle's four causes from *Physics* II:3 and *Metaphysics* V:2.

p. 45 §3 The *formal* cause is the shape, form, or completion that gives the mode and intrinsic and appropriate being to that composite, which makes it be what it is called: for example, the proportions of lines.

p. 45 §4 The *efficient* cause is the cause from which comes the principle of that movement and action necessary for the production of its effect: for example, in the proportion of the straight line, the man who made it is the efficient cause.

p. 45 §5 The *final* cause is the usefulness or apparent good, for which every agent is induced and moved to do its actions so as not to act in vain: as can be seen in the example of the straight thrust: the apparent good of achieving victory through that thrust is the final cause.

p. 45 §6 Thus, all composite things, both artificial and natural, depend on matter
p. 46 §1 and form as intrinsic causes and essential parts; similarly they depend on the artificer who makes them and the end that moves to make them as extrinsic and foreign causes in their production.

p. 46 §2 In any sort of cause or its mode, both the cause and the effect can be considered. This can sometimes be in potential (readiness to be made) and sometimes in actuality (being produced). For example, we can correctly say that a man, before he makes a thrust, is potentially the efficient cause of it (that is, in readiness and aptitude to be able to do it). But we would not say that the man is actually the cause of the said thrust until he has actually produced it.

p. 46 §3 Measure is that by which we ascertain the quantity of something, and that thing is the amount of ground which is between the two combatants. Until now, we have sought knowledge of it more or less through practice. Thus many are used to measuring themselves when they show off (as the Spaniards do): which is uncertain.

p. 46 §4 But we, wanting to establish a certain and determinable measure, will
p. 47 §1 consider the second circle, which forms from a body standing in the first circle (as was discussed earlier).[1]

p. 47 §2 For although men are unequal in stature, they none the less will describe the second circle properly in accordance with their size. For even assuming that the enemy is taller, he's never so large that he can overcome their opponent's advantage (who is in the first [stationary] circle.)

1 p. 17 §8–9 "The first of the two is formed when, standing balanced, a circle is described around their feet" & "The second is when, standing in said circle, a circle is described at a distance from their body that is equal to their height".

p. 47 §3 We therefore conclude that for the enemy to reach the centre of the said circle, he must necessarily arrive with one of their feet upon the circumference. That's because with the length of the sword — which is two arms' lengths, and with all that is occupied by the hand and with the arm itself, forms exactly three arms' lengths — it's the exact distance from the circumference to the centre.[1]

p. 47 §4 Thus, this person who finds themself in position can open their compass and, striking, can reach the distant enemy from the circumference by as much as that opening of their compass: as can be seen in this figure:

1 In other words, a person's reach with a rapier is the same distance from their front foot as their height.

p. 49 §1 Timing has such a close relationship and connection with movement, from which it is never separated (since there can be no movement that is not either slow or fast, and consequently made in more or less time), that it is necessary for it to be substantially or accidentally linked with movement; that is, it must either be the same thing as movement or an intrinsic accident of it.

p. 49 §2 Hence, it is clear that if timing were movement, it would follow that, just as speed and slowness apply to one, they would also apply to the other: because, although it is true that we can say that this or that movement is fast or slow, nevertheless, we wouldn't call any timing fast or slow, as nothing can define itself by itself; thus, it remains that at least it must be an accident of it.

p. 49 §3 Because those accidents, with their quantity, make known and determine the quantity of some subject, they can for this reason be called its measure.

p. 49 §4 Timing, therefore, is the number or measure of movement, according to
which two instants — one before and the other after — determine the move-
p. 50 §1 ment from both sides. For there are three timings: past, present, and future,
among which it is not necessary for us to speak of the past.

p. 50 §2 The present timing in the actions of arms cannot be known by us except accidentally, when it happens that the opponent acts according to the usual manner of others of whom we've knowledge, because if timing is the measure of motion, we can't have knowledge of such timing unless we first consider the nature of the motion.

p. 50 §3 For such a motion, when it is done by one person of their own will, and the other must receive the motion from that person, this one can't, at the same time that the other begins the motion, also put themselves in motion; but they can indeed do so afterward. Consequently, the one who is the first to start the motion will finish before the other, who must receive the motion from them. Therefore, regarding this sort of measure of motion which is called fixed timing, we can't have knowledge.

p. 50 §4 Therefore, in order for us to have knowledge of timing, we must operate in two
ways. In one, since motion arises from stillness, we will consider the nature of
p. 51 §1 the stillness of the enemy's sword and equally the stillness of the enemy's stance,
because these will reveal to us the future motion, and consequently, we'll have
knowledge of its measure — that is, of the future timing, called premeditated.

p. 51 §2 Additionally, we will initiate the motion for the enemy, thereby compelling them to make another motion; consequently, we'll have knowledge of the future timing.

p. 51 §3 We'll note, however, that the motion we make to provoke the opponent should be brief, so it concludes before the enemy becomes aware of it, and that, if they intend to strike, they'll need to make a greater motion than the one we made to initiate theirs. This will be easy for us if we keep the body unified, allowing it to immediately obey our will.

p. 51 §4 Thus, in order to distinctly recognize timing, we will apply the timings revealed by the measure of singing, or music, to the movements we make with our steps.

Since we might find ourselves with the left foot forward in a stance, when we move from that position and counter-pass with the right foot, reaching the fourth circumference, the measure of such motion will be *eight beats,* which equates to a *Maxima.*[1]

p. 51 §5 With the right foot forward, we can then make four movements, as we've said:

p. 52 §1
1. The first is when, finding ourselves in the first circle with the right foot in motion, we make the forced step and reach the fourth circumference. This motion takes *four beats,* like a *long note.*[2] We'll only use this motion to keep the enemy away.

p. 52 §2
2. When we then make the forceful step, described as an equilateral triangle (which we use to execute a thrust), the measure of this movement consists of *two beats,* equivalent to a *breve.*

p. 52 §3
3. The measure of the step, formed when the right foot touches the second circumference while in motion, is *one beat,* equivalent to a *semibreve.*

p. 52 §4
4. In making a half step, we will measure a movement that amounts to *half a beat,* the same as the duration of a *minim.*

p. 52 §5
5. That slight compass opening, made when the body is in the second stillness, measures the time equivalent to a *semiminim,* with *two of these* together equating to *one minim.*

p. 52 §6 Similarly, to provide some rule for recognizing the time to strike; I state that it occurs when the opponent makes any movement, whether of the sword alone or with one of the bases, be it the right or the left, and when the sword is used to make feints and provocations. We will always strike in those timings, whether they are thrusts, withdrawals, rotations, or transports.

p. 52 §7 And there are three ways to strike: *ahead-of-that-timing, within-that-timing,* and *after-that-timing.*

---

1 Dance was performed to music with two four-beat bars, making an eight-beat section. A *maxima* is an *'octuple whole note'* in modern music theory.

2 A *'quadruple whole note'* in modern music theory.

- p. 52 §8 Striking *ahead-of-that-timing* is called striking presumptively; it occurs
  p. 53 §1 when the enemy is in a guarded position with their sword. In this case, the strike is directed at the nearest part.
- p. 53 §2 Striking *within-that-timing* occurs when the enemy is in motion, either executing an attack or parrying; in this case, you strike the part that is moving, which will be the right side if it is an attack, or the left side if it is a defensive move.
- p. 53 §3 Striking *after-that-timing* happens when the motion is about to end, returning to stillness; you follow the trajectory of the attack. It is time to strike when the sword exits the imaginary straight line passing through our line of sight.

p. 53 §4 All wise men praise things done in their proper time and place, and never outside of it. Therefore, it is to be considered that just as it is fitting to wait for and choose the right timing to act, so it is essential to be diligent and not let the precise moment, in which it's appropriate to do what we intend, slip away. This is what we call an opportunity or conjunction, which, once lost, is rarely reacquired.

p. 53 §5 This opportunity is a part of timing that carries with it the appropriateness
p. 54 §1 of doing something. This is considered in two ways: first, when the opponent provokes us with some attack, or makes feints or invitations, or grants us the straight line; in such a case, one should not miss the opportunity to strike the enemy.

p. 54 §2 The second way to seize an opportunity is when we create it ourselves, but by appropriate means and not by improper measures; these are errors, as will become apparent from what follows.

p. 54 §3 Place is nothing other than the surface inside and the limit of that body which encompasses it, touching and fitting closely on every side with the outermost surface of the contained body.

p. 54 §4 Thus, in our context, the surface of the opposing body will be that which the outermost surface of the sword touches on every side when it strikes.

p. 54 §5 Because in this matter it is customary to say that something is found in a place in two ways: that is, in a common place and in a proper place. Therefore, the common place of the strike will be the entire surface of the whole body; the proper place will be a part of the surface, which is usually the part that lacks defence and protrudes the most.

p. 55 §1 Since the centre of the body in all its movements is the part that remains the most steady, we'll strike that as the proper place.

p. 55 §2 Since, as we have seen in the composition of man, our body is composed of the four elements. Of these, the two heavy ones cause a weight in us, that is the power that tends downward, and also that resisting contrary motion — that is, for anyone who'd wish to pull it upwards — and since the members of our body serve our will as instruments of the motive force, we would not be able to obey with the said members unless the weight is distributed among them as the need will require.

p. 55 §3 We must therefore understand that the body's weight is distributed over the
p. 56 §1 two legs, and it can be shifted and unloaded now onto one, now onto the other. But this must be done with care: since these two legs are the ends of the perpendicular, which is in their middle, and it's impossible to move from one end to the other without passing through the middle, it will be necessary for the weight to be transferred onto the legs using these two stops. For example, if the weight is on the left leg and we want to transfer it to the right, it will be necessary to first rest it in the middle — that is, on the perpendicular — and then unload it onto the right. This will be similarly done from the right to the left.

p. 56 §2 Since the part of the body that is to make any motion, whether straight or angled, would not be able to do so comfortably unless it's relieved of part of its weight, the leg that is to remain still, as if it were the leg of a compass, will always bear all the weight so that the other leg, being lighter, can obey the will and execute the commanded motion.

p. 56 §3 When walking naturally with deliberate steps, the weight will be kept united in action so that it can be placed on one of the two legs as needed.

p. 56 §4 Similarly, if a person wants to gain ground to the right or left side, they will have to move by placing the weight in the middle and then unloading it onto the right leg, pulling the left leg into the first circle. By proceeding in this manner, they will always be ready to respond to an attack. Therefore, one should never make a continuous movement with both legs in a narrow space, except when walking.

p. 56 §5 So in order for a person to achieve these particularities, we must know that
p. 57 §1 the second circle,[1] which they form in motion, is divided into four right angles by two diameters. When standing in the first circle, that is, balanced, and needing to move either to the right, to the left, forward, or backward, they will always do so by moving only one of the legs, keeping the other stable in the centre of the first circle.

1 "[The second circle of motion is when] he stabilises the left foot and describes another circumference with the mobile one: in this, we enter by sideways movement to exit with thrusts." p. 19 §2.

p. 57 §2 It is to be noted that while we find ourselves with our feet in the first circle, and exit with one leg from the first circle to enter the second circle, which is done in motion, we will then divide that second circle into four planar figures, each trilateral, enclosed, and contained by three straight lines. When considering each triangle separately, every time a person finds themselves positioned in one of them and wants to enter another, they must unite and return to the first circle made in a stationary stance; then, as needed, they will move into one of the said four figures. This is all demonstrated by the following figure:

p. 59 §1 *Having explained the seven circumstances that are involved in the use of arms, we must know that, in order for any of our actions to depend on virtue, it is necessary that it possesses four conditions – that it be spontaneous, considered, chosen, and willed.*

p. 59 §2 An action is understood to be spontaneous when we perform it by our own will; Conversely, those operations that are not performed willingly can be called non-spontaneous.

p. 59 §3 From this, we derive a rule for ourselves: that in the actions of arms, a person should perform all their operations voluntarily, being the first to commit resolutely; consequently, they will act in a way that forces their opponent to do everything against their will, by force requiring them to parry.

p. 59 §4 Consideration, as defined by Aristotle, concerns those things which can fall under human counsel.

p. 59 §5 Therefore, in this art of ours, we will thoroughly consider all the means that can lead us to our end, which is victory.

p. 60 §1 Choice is nothing more than a considered consent or assent to those things that are placed within us: for since a thing is first consultable, and then choosable, if it is first considered and judged to be good, it will reasonably be chosen.

p. 60 §2 Thus, from these two conditions, which I have joined together, we will establish a rule: that after having considered the means of offence and defence, we will choose the best proportion of attack and defence, and we will continuously operate with that in mind.

p. 60 §3 Our will is ready to regard that end, which is not only a true good but also an apparent one, for the acquisition of which we must then spontaneously consult within ourselves the means that can lead us to that end; and those, finally, choosing virtuously to operate as we will do in the contest of arms, which is done body to body; if, having regard to the true good, which in this action is to defend oneself and to offend the enemy, we will courageously execute the things considered and chosen; otherwise, by doing so, we nullify the entire art of arms.

p. 60 §4 So, given that from one point to another there is only a straight line, and
p. 61 §1 given that when a line is situated in a body, another cannot enter it (because two bodies can't be in the same subject at the same time), it follows that every time two come to fight each other, the circle of distance will necessarily be born (which is divided by its diameter, and each of them forming a point). Whomever of them first places their sword in the diameter (which consequently becomes the straight line) will force the opponent to pass through it, such that they will have both the offence and the defence, since the enemy's sword becomes parallel to their body. To better understand this, one must carefully consider the following figures, below:

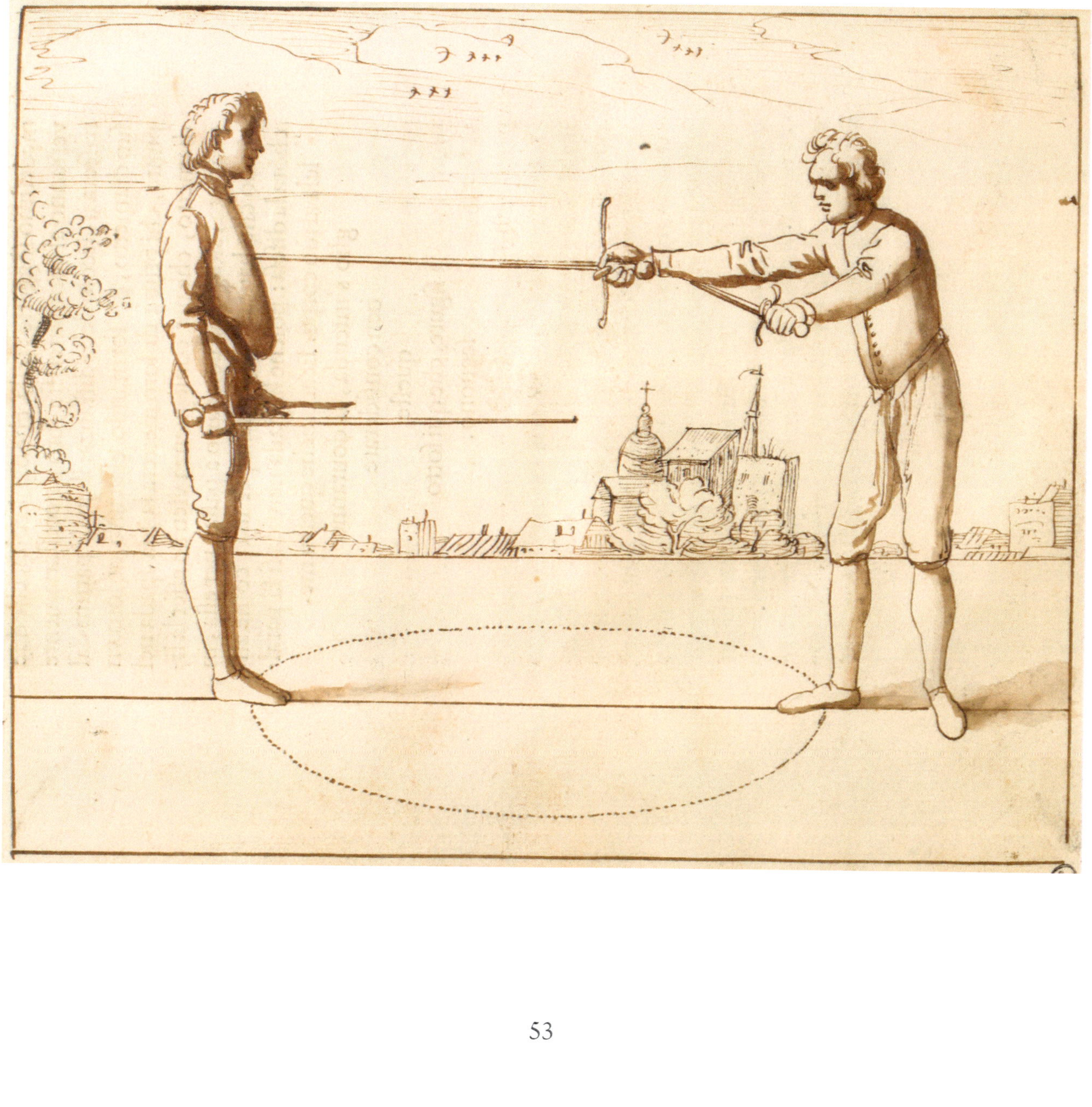

p. 63 §1 This is the foundation of the science of arms. Hence there have been various operations by those who have dealt with such matters so far. Either they've placed themselves in it [the straight line], or have yielded the place – so that by the enemy striking they could secure it for themselves in that moment, by parrying, and by placing their sword in the straight line, they could then strike.

p. 63 §2 Those who have placed the sword forward in a straight line have done so for the reason stated, which is very important; to keep themselves covered by the sword, and to keep the enemy at a distance. From this have arisen many and various different ways to gain the advantage.

p. 63 §3 So first, according to what is customary in Spain, and also in Italy, the approach is to gain control by dividing the sword into three parts: of these, the first part near the hilt is the strongest because it is closest to the driving force; and the last part is weak, being far from the driving force. But it should be noted that I say weak in parrying, although it is very strong in striking.

p. 63 §4 Proceeding with this division, one should aim to place their sword above the
p. 64 §1 enemy's at their weak part, keeping the arm close to the body, and the hand's hilt positioned at the same level, or slightly lower than the enemy's. This way, while waiting for the opponent to release, at the same time one can gain the straight line by extending the arm. One should stick to a rule that, one must move towards the sword at one time, and with another, towards the enemy's body.

p. 64 §2 To defend against it, it used to be that, while the opponent's sword was in motion, one would disengage in said motion and free the sword in order to overcome it, either by simply moving the sword or by using the technique of "shaking" (which involves shifting the body from one foot to the other, or by retreating the left foot backward). Thus, proceeding step-by-step, one would engage against the other. However, this is more useful for playing around than for serious combat, given the nature of men (who are unequal in size and strength), as well as swords (being unequal in length and weight). Moreover, an agitated person does not notice these subtleties, and since this technique is based on uncertain principles, it will also be uncertain.

p. 64 §3 Some remedy many of these inconveniences, should they find themselves in said line with their sword, like so: when the enemy goes to gain it, at that moment they move their sword out of the straight line, and encountering the enemy's sword out of force, then placing the left shoulder without disengaging their sword from the opponent's in that angle formed by the two swords, they step across with the left foot and strike with a backhand thrust.

p. 65 §1 Many, having considered that while gaining the weak part of the enemy's sword they could be injured by the opponent withdrawing their sword, have followed this order in approaching the enemy's sword: to move the body towards it, so that if the opponent disengaged and struck, they would find no place to strike. Hence, they in that movement placed their sword in a direct line to strike.

p. 65 §2 But when the enemy's sword was high, they provided their opponent with the opportunity to strike. They did this with certain cuts of the sword accompanied and united with the arm, and with the same movement of the body to the right or left. As the enemy struck, they in that moment would bind the enemy's sword with a whirl,[1] and strike with the dagger; or they would strike by placing their sword in a direct line at that moment.

p. 65 §3 Likewise, it has been customary to make feints to distract their position off
p. 66 §1 this straight line — either inside or outside; with thrusts or cuts; to intimidate the one who held the sword in front. Thus, by leaving the line with their sword to parry, they would give up space for the enemy to place their sword. But those who make these feints first commit an error (to cause the opponent to make another error); these errors in the acts of arms are then paid for with the loss of life.

p. 66 §2 Since someone in this posture will always maintain that advantage, if they keep the point of their sword fixed on the enemy's right shoulder with every small movement, thus we follow it with the posture wherever it goes. Consequently, many try to find the sword, not by pressing its weak part, but by overshadowing it with their own. So that they would have the distance to strike, they do this with the left foot forward, and they are careful to place the left foot exactly in line with the enemy's foot, and the sword extended in a straight line level with the enemy's. When they reach the appropriate distance, they strike by advancing with the right foot.

p. 66 §3 However, the person who finds themselves in this line will strike the opponent if they make a sideways movement with the right foot. When the opponent is provoked from the outside, they can strike with a subjugated thrust[2] (as it's called by these fencing masters, also called a "*hunch*");[3] or when provoked from the inside with a backhand thrust (or a "*header*").[1]

---

1 *un molinello* – literally a winch or a mill, perhaps like *moulinet*, a cut around.
2 *potrà ferire di punta scavizzata* – *schiavizzata* means 'enslaved'.
3 *gobba* – hunch, hump or hunchback.
4 *incapocchiata* – Florio's dictionary gives 'to enhood or encowl' for *incapucchiare*.

p. 66 §4 Others have based their approach on attacks and defences and have proceeded
p. 67 §1 in various ways; either by attempting some kind of cut or thrust to provoke the enemy while they are in guard, so that by disrupting the enemy either with a parry or a strike, they can then parry and subsequently strike.

p. 67 §2 Or, when they have found the sword in presence, they have disrupted it either by striking it downwards on its weak part with the true edge, or by lifting it up with the false edge from below, so that they could strike during that timing: in which way, parries and strikes become endless — as can be seen in the books that have dealt with this subject.

p. 67 §3 Those who expose an open target do so to give the enemy the opportunity to strike, so that they can then parry and strike back, but feints greatly disadvantage these practitioners; and beyond this the same approach becomes endless.

p. 67 §4 But in order to avoid these kinds of attacks and defences, we will neither
p. 68 §1 concern ourselves with putting our sword in a straight line, nor will we try to gain it with the aforementioned methods if the enemy's sword is there; rather, we will act according to due reason. Since true knowledge consists in understanding the causes of things, once these causes are understood and removed, the effect is also removed. We will consider the cause of such a line, which is the point of the enemy's shoulder and the point of our body where this line aims. Every time the point of this line is removed from our body's edge we remove the consequent effect of the line.

p. 68 §2 This point [of the line] can be removed [from our body's edge] in two ways:

- In one, without losing ground, by moving sideways either to the right hand or to the left hand.
- In the other, by losing ground by retreating a step and striking at the same time.

p. 68 §3 Moving the point out of the straight line is done in this way — that is, having divided the body into two equal parts by the diameter of our circle, in each part we establish a point. Either we place one of these points voluntarily at the outside of the line, or the enemy is the one who places their sword to one of them. When we place it of our own will, it will always be the right side, to draw out the wound more easily and conveniently, because in this way we will gain a step of ground.

p. 68 §4 When the enemy places their sword towards one of our two points, if by chance they place it to the right, we will move out with the left foot; and if to the left, we will move out with the right foot. By doing so, the effect of the enemy's line is removed.

p. 68 §5 Moving the point, in addition to nullifying the enemy's line, generates
p. 69 §1 another line for us, into which we can place our sword to wound; and thus it saves our body, which avoids the initial wounds, and furthermore, it secures us from the subsequent ones.

p. 69 §2 When the point is moved by losing ground, that is, when we move it away from the enemy's sword, we will note that this is done only by striking the enemy: when it happens that they attack our right side we will then withdraw it with a step back; and at the same time, we will strike their arm. But if they strike our left side, we will lose ground with the left foot, and in that timing, we will strike. However, we do not use this method, since we can operate in a better manner.

p. 69 §3 Since every kind of cause and every kind of effect can sometimes be considered in potentiality for production, and other times in actuality; thus we will consider that line either in actuality, which will be when the sword is already placed in it; or in potentiality, when it's not there yet, but is well-suited to reach it. Considering that all lines, in whatever sort of proportion they may be, ultimately end up in the straight line when wounding, as can be seen in the following figures:

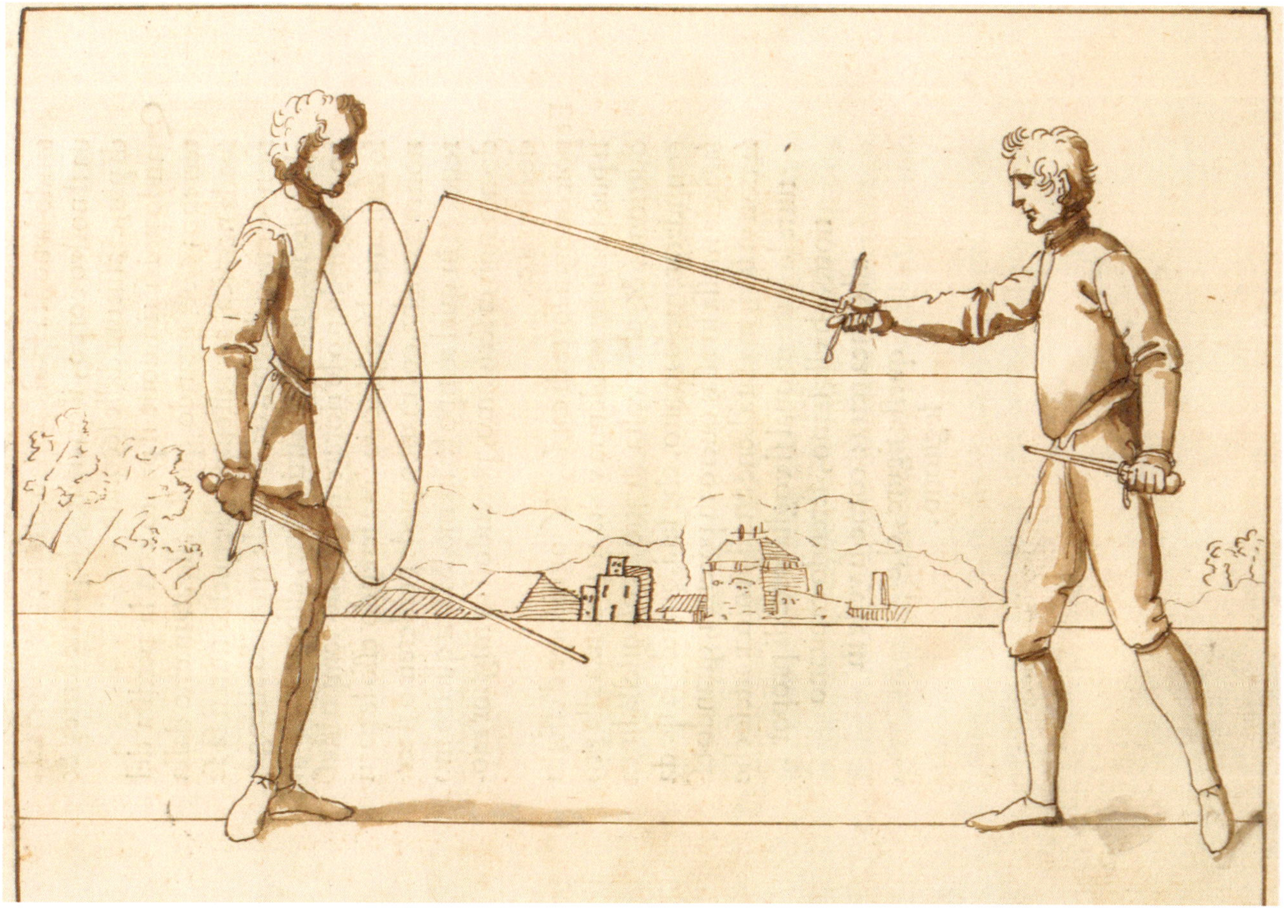

p. 71 §1 Therefore we'll distinguish between our opponent either being in a stance (that is, still); or departing from it, being in motion.

p. 71 §2 While they are still, they will either hold their sword in a straight line (that is, in the very act of production), or if it is outside of said line they will hold it in potentiality (that is, ready to place it into the line).

p. 71 §3 When they hold the sword in the very act of production (that is, in the straight line), we should consider that, according to the opening of their stance, by lowering their right shoulder, the sword will go level with the shoulder. Consequently, the point of our surface will be directly in line with that line, which will always be called a straight line as long as in said proportion it reaches the direct centre of our body.

p. 71 §4 While our enemy holds the sword in potentiality to place it in the straight line, it will be called an angled line; which will either be forward (in presence), or backward (out of presence).

p. 71 §5 The sword can be said to be angled in presence in three ways:

p. 72 §1

1. First, when it is indeed in presence in the diameter of our circle but is found below the centre of our body.
2. Secondly, the enemy's sword is said to be angled from our body when both are in the diameter of the circle of distance, and the point of the enemy's sword is outside our body (either high, low, or in any kind of proportion), and this will be when the enemy holds the hilt of the sword at our right or left parts.
3. Thirdly, the enemy's sword is angled from our body when it forms an angle with the arm and its point is above the centre of our body.

p. 72 §2 Out of presence, or backward, the enemy can hold the sword in two ways: either with the right foot forward or the left:

- With the right foot forward, he will hold it either withdrawn and low, or high and in the manner of doing a cut.
- When he holds it with the left foot forward, he will either hold it high in the guard of the falcon[1] or low in a guard called by many the long and wide tail.[2]

1 This term appears as early as ms. 01020 (Q2 C15th), and is depicted in Vadi's longsword (1480s). It's used in ms. Ricc.2541 (Q3 C16th) in the same manner as Ghislieri for a high 1st position. The jargon was also referenced derisively by Viggiani (1530s).

2 *coda lunga e larga* – a term used by Marozzo (1536).

p. 72 §3 And all these lines can be placed in infinite proportions, and from this, it has come about that so many guards have been formed. For postures can be formed from all the principles of the strikes; from all their means and from all their ends. However, so that we don't confuse ourselves, we'll only pay attention to those proportions that we've mentioned, in which the enemy will inevitably find themselves regardless of whether they're a little higher or a little lower.

p. 73 §1 Now, returning to the explanation of our art, I say that when the enemy is found in any kind of still posture, to operate with knowledge we must consider the four causes that produce the effects:

1. Firstly, the efficient cause, which is the man themself. This is the most general and most remote cause in the action of this posture. However, the most immediate and particular cause will be the act in which this man finds themself. Which act that is shows us the effect that can arise from it.

p. 73 §2
2. The formal cause, which is the general movement from their stillness. This indicates to us the measure of the particular movement being put into effect.

p. 73 §3
3. The efficient cause,[1] which is the proportion of the enemy's sword. This reveals to us the effect of its strike.

p. 73 §4
4. The final cause I now wish to consider in this effect is that which we consider about the line of the sword, which, when striking, aims to hit our body. Therefore, the final cause will be the point of our surface.

p. 73 §5 Thus, when the enemy has their sword in a straight line but positioned below
p. 74 §1 our dagger, as if they were in 3rd or 4th, then by striking that line and simultaneously thrusting, we will place our sword in the straight line; but with only the sword, we will step out with the left foot while striking.

p. 74 §2 When the enemy holds their sword against us in a straight line — so either with their left foot forward (as in the guard of the head, or of entering, or of the face), or with their right foot forward (as many do in 1st, 2nd, and 3rd) — then we'll take as a rule to step out of that straight line with a sideways motion.

p. 74 §3 Then, when the enemy's sword is angled from our body (that is, in its first way, below the centre of our body), then we must strike with the left foot during that stroke, to form the evasion of the body. But here I speak of using the sword alone — because if we have the dagger, we will place it on the imagined line and strike in our usual manner.

1 The original repeats 'La causa formale,' twice, probably in error. I've corrected it based on the lists of causes on p. 45.

p. 74 §4 If the enemy has their sword angled from our body (in the second manner — in this position, we will see the body of the sword but not its point), then we will always move to our right hand side, avoiding that length. Thus, thrusting, we will place our sword in the diagonal.

p. 74 §5 But if the sword is angled in the third manner, we must distinguish that if the
p. 75 §1 point of that angular line is directly opposite our right point, then we will act against it as if it were in a straight line. But if the point of the sword is opposite our left point, we will block it, thrusting our sword in a straight line into the attack.

p. 75 §2 In the case that our opponent holds their sword back or out of presence with their right foot forward, (as some do when they hold the sword drawn back to perform a '4ther'),[1] or when they are in a high guard in long and wide tail, or with the left foot forward in falcon guard, or in long and high tail,[2] and finally in any sort of proportion as long as it is out of presence, we will always strive to engage the body, and seek to move it with our right parts, since it is natural to all animals – as Aristotle says.[3]

---

1 *inquartata* – 'en-fourth-ed', referring to Agrippa's guard of 4th.

2 '*in guardia alta in coda lunga, & larga; overo col piede manco avanti in guardia di falcone; overo in coda lunga, & alta:*' Marozzo (1536) again. There's a reason that Agrippa's numbered guard names replaced the older metaphors.

3 Referring to Aristotle's *On the Gait of Animals*, Part 4.

p. 75 §3 *Now that we've explained the proportions of the lines when they're at rest, we shall discuss those that are in motion.*

p. 75 §4 We must therefore consider that they originate from a state of rest, and come to rest in another: for all attacks originate from the centre of the shoulder, which is the origin of the straight line, and in the attack, they return again to that centre.

p. 75 §5 So, because all movements of a man striking are mixed (for example, to to
p. 76 §1 execute a thrust it's first necessary to make a circular movement to cause the straight one); and likewise because a man forming a strike does so with both violent and natural movements (though the violent one is more vehement than the natural one), we'll distinguish that movement has a beginning, middle, and end.

p. 76 §2 So that this work can be better understood, in this following part I'll set aside the angular and forceful movement that a person makes when preparing the arm, lifting it, and resting it in a position to form either a cut or a thrust, even though that's time to strike the enemy. Nevertheless, I intend only to speak of the movement that occurs when the arm, departing from that state of rest, descends in motion toward the strike. For this movement is (as we've said) composed of both forceful and natural elements: forceful with respect to the force of the moving power, which pushes and forces the weight downward; natural with respect to the weight of the arm and the sword, which naturally tends toward the centre of the earth.

p. 76 §3 Hence, that movement has its weak beginning, which is when it starts to
p. 77 §1 descend; it then has its strong middle, which is when the arm reaches the level of the shoulder: at this point is the greatest blow a man can deliver, because here ends the effort of man's force; finally, it has its end, force spent or less strong, which is when the arm exceeds the plane of the shoulder: because the movement becomes more natural, as the effort ceases, the arm being scarcely assisted by the remaining motive power.

p. 77 §2 By distinguishing these three terms differently, we must operate against each one of them individually. Parrying should be avoided because it's the worst option with a sword. That's because (unless nature concedes to vacuums)[1] – if one is the first to strike, and the other parries, the one who's first to strike can always keep the other parrying by continuing their motion. Hence, we'll defend ourselves from enemy attacks in two ways.

1 *"perche non si concedendo il vacuo nella natura"* referring to Aristotle's *Physics*, Book IV, section 8: *"In a void, no one could say why a thing once set in motion should stop anywhere; for why should it stop*

p. 77 §3 In the first, if we find ourselves at the distance, we can strike at the beginning of the movement: thus, we'll interrupt the opponent, who, once hit (due to a natural aversion to harm), won't continue the movement he started. This will be called striking *ahead-of-that-timing*, while the movement is weak.

p. 77 §4 Thus, we'll take great care to defend ourselves resolutely in this manner, positioning the body well in profile, extending oneself, and striking at the enemy's right parts: this should be done until the weight of the arm, or the mobile leg in that movement, is pointed down to the ground.

p. 78 §1 Moreover, swords that are sharp and pointed have a different effect from foils.[1] For if we were to strike our opponent, with a sword that thrusts and penetrates, when their weight is already going downwards, we'd indeed injure them, but the enemy could still strike us.

p. 78 §2 The reason for this is that if the body's weight, which was balanced with its diameter perpendicular to the flat surface of the ground, is moved from that vertical position – its point of equilibrium (i.e., from its state of rest), and that weight is in descent towards its natural centre, it can't rise back up. This is because all things move with difficulty towards the opposite of their intrinsic inclination.

p. 78 §3 To confirm this, consider that the human soul, being just one, can't be greatly focused on many things at once: that is, to command movements of the hands and other limbs with reason, and to perceive the pain of wounds with the sensory faculties at the same time.

p. 78 §4 Therefore, we conclude that it's best to strike the enemy before their body weight is falling downwards.

---

here rather than here? So that a thing will either be at rest or must be moved ad infinitum, unless something more powerful gets in its way.

Further, things are now thought to move into the void because it yields; but in a void this quality is present equally everywhere, so that things should move in all directions."

The idea was restated as *"Natura abhorret vacuum"* by François Rabelais in the 1530s which also described fencing practice; see *'Gargantua and Pantagruel'*, by François Rabelais, trans. M. A. Screech, Penguin Books, 2006, found via Behringer, W. (2009); and 'Arena and Pall Mall: Sport in the Early Modern Period.' *German History*, 27(3), 331-357. doi:10.1093/gerhis/ghp03

The theory was supported and restated by Ghislieri's acquaintance Galileo in the early C17th as *"Resistenza del vacuo"*. René Dugas (1988). *A history of mechanics*, Courier Dover Publications. p. 144.

1 *marra* – a term used by contemporary fencing authors to refer to safe training tools, such as Viggiani and Altoni. *"Spada da marra"* is referenced in the *Vocabolario degli Accademici della Crusca* when differentiating a sharp sword from a fencing sword. p. 348 of the 1612 edition (**FILO**). Thanks to Jherek Swagner for bringing this to my attention.

p. 78 §5 The second method of defending ourselves against moving lines can be done in two ways.

p. 79 §1

1. In the first, we can give ground – withdrawing the point of our body in one of the two methods previously mentioned, when losing a step of ground in order to move out of the line with the points; and striking the opponent's arm. Or, we can counter-step with the right, moving out of the point of presence, and still strike the arm with a chop down.[1] This is called striking *within-that-timing,* during the strong motion.

p. 79 §2

2. In the other method, we let the attacks pass into empty space; and we strike with the types of strikes that suit us best, as soon as the sword passes through the rays of our vision. That is, striking *after-that-timing,* during the weak motion.

---

1 *Stramazzone* – literally *'over-macing'*, used by other fencing authors such as Vadi (1480s) and Capoferro (1610), who defines it as cut made with a turn of the wrist on p. 40.

p. 79 §3 *Yet most men don't have this subtle knowledge of how to attack others, and defend themselves. Men driven by nature tend to parry attacks in the middle of their motion; that is, in the strong part, when the arm reaches the level of the shoulder. Therefore, in discussing parries, we shall choose the best one*[1]

p. 79 §4 So, there are four ways to parry:

p. 80 §1 1. In the first, all movements are aided in their descent, because everything, while in motion, moves more easily than if it were to suddenly depart from rest. This is executed in the following manner: step out of the straight line with the right foot and, with the left foot accompanied by a backhand across, assist any attack coming from the enemy's right side in its descent. Then, returning in with the right foot, strike the part closest to us.

p. 80 §2 Thus, in all attacks coming from the left side of the enemy (such as all types of backhands and those composed of them) stepping out of the diameter of the circle towards the enemy's left side, we'll assist those lines in their descent with a cut across, and with the least motion, we'll more easily strike.

p. 80 §3 2. In the second way of parrying, we counter the enemy's strike, which is disrupted in this manner (and this method is good for making the enemy drop their sword): as soon as the enemy makes any attack (whether a cut or a thrust) coming from the right side, gain ground with the right foot followed by the left in a natural motion towards the enemy's left side: and with a half-forehand, cut that line in its weak part (that is, in the last third towards the tip of the sword) and respond with the most convenient strike.

p. 80 §4 Similarly, all attacks coming from the left side towards the ground with
p. 81 §1 an advance of the left foot are cut with a half-backhand, so the enemy is struck by advancing the right foot.[2]

1 I believe this section draws on dall'Agocchie (1572).

2 The author breaks up the list of four parries, with the subsequent section describing two additional disarms, and then returns to the parries. I've taken the liberty of re-arranging the text.

p. 82 §2 3. In the third method of parrying, one resists all attacks with the true edge. Although it's possible to do so with the false edge as well, we omit it due to its weakness. By positioning the sword in a straight line, one parries and strikes by making the edges of the swords slide against each other. For all proportions of lines made from the enemy's right side, one parries with the guard of entering, either with the left foot or the right foot forward. For those coming from the enemy's left side, one parries by turning the true edge in the guard of facing.[1]

p. 82 §3 4. In the fourth and final method, one parries all attacks in the guard of the head.[2] This method is more secure than the third because it's not susceptible to feints. One responds with the most suitable strike, taking care to receive the blow on the first third of the sword, the strongest part. Then, by gathering the left foot close to the right in the parry, one can strike with either a forehand cut or a backhand cut.

p. 82 §4 When someone is strong, they can parry effectively by keeping the sword slightly angled in a strong position and then strike with the most convenient attack following the parry.

---

1 These guards are given by Marozzo in *Opera Nova* (1536), but Ghislieri is probably drawing from dall'Agocchie's *Dell'Arte di Scrima Libri Tre* (1572).

2 Marozzo gives this guard as the arm extended at the opponent, but now the blade pointing upwards. Dall'Agocchie's version is pointing down.

p. 81 §2 *And indeed, we should say that in addition to the aforementioned method (since it's relevant), there are two other ways to make the enemy drop their sword.*

p. 81 §3 In the first method, using only the sword, one employs the "Squiggle"[1] (which is what this technique is called). This is executed with both a forehand cut and a backhand cut. With these two cuts, one should aim to take the enemy's sword at its final third, positioning the hilt of our sword lower than the enemy's point and striking the edge of the enemy's sword with a twisting motion of the hand, causing it to drop. If one wants to attempt this with a backhand cut, the hilt should be at the same level, differing only in that the point of our sword will aim at the enemy's left side, using a backhanded whirling cut to strike the false edge of the enemy's sword, thereby easily achieving our aim.

p. 81 §4 In the second method, one places our sword beneath the enemy's sword, and
p. 82 §1 the point of our sword rests on the enemy's right arm. Then, with our dagger positioned inside, we turn outward, performing the action of a lever, thus causing the enemy to drop their sword.

1 *Garatusa* – could mean '*wantonness*' (see *A Dictionary, Spanish and English*, G. Baretti, F. Wingrave, 1800), otherwise a '*scribble*', a '*caress*', or a Spanish card game where one discards their cards to win. The name is used in Godinho (1599) for an envelopment of the opponent's blade when it's held in 4th. The term also appears in Carranza (1582), Pacheco, Cruzado and Mendoza, Pedro de Heredia, and a $C18^{th}$ Spanish dictionary (*Academia Autoridades (G-M)*, 1734).

p. 82 §5 *For all these types of line positions, both when stationary and in motion, it's necessary*
p. 83 §1 *to position the body in four specific stances, which we'll call guards. This is not to wait for the enemy but to engage them and perform both offence and defence. Therefore, I'll establish four guards corresponding to our four types of attacks and four methods of action, each illustrated with its respective attack.*[1]

p. 83 §2 The first one of these is shown below, in which we've adopted the initial posture of the body as devised by Polykleitos, which places all the weight on the rear leg.[2] This is to ensure the body can respond to the will.

p. 83 §3 The right foot is lighter because it will be the one to move, and the body is positioned in the first of the four circles that form in motion, allowing it to quickly pivot in all directions.

p. 83 §4 The action is with the right foot forward, as Vegetius describes in Chapter
p. 84 §1 20.[3] It was the custom of the ancient Romans to have the left foot forward to increase the force of the throw when throwing their javelins. But when using the sword, they kept the right foot forward to cover their width and to bring the sword closer to the enemy's point.

p. 84 §2 Of the three proportions of the body in this posture, we adopt the foreshortened one (that is, between facing and profile).[4] This is taken to show the enemy the smallest surface possible, thus denying the enemy access to all but one point. To move away the enemy's sword, the body is bent forward with the initial movement, thus bringing the sword closer to the opponent's point.

p. 84 §3 Since the sword, while in position, indicates the future strike, it must be positioned so that it can potentially deliver any type of attack without being limited by the hand's posture. The posture should be such that, with the body foreshortened and the arm raised naturally, the sword and arm form a single line with the body. Thus, drawing a line from the tip of the sword forms a square figure, as shown by the posture.

p. 84 §4 In this position, the sword is close to the enemy's point because simply turning the hand, the sword forms an equilateral triangle in the motion. With it, the sword will enter in a straight line while it strikes.

---

1 I've taken the liberty of turning this bridging paragraph into a heading.

2 See p. 17 §3.

3 Referring to the chapter on *"Troops in Action."*

4 The description of *scurzo* here doesn't translate well literally, but you get the gist.

p. 84 §5 In its descending attacks, from the circumference to the centre of natural
p. 85 §1 motion (from high to low), the sword will search the whole body — which is the common area of the strike. It will form a semicircle in its motion, and because the hand forms an equilateral triangle, the sword presents three views to the enemy's eye.

p. 85 §2 With this posture, the person strikes using the fourth length,[1] which is superior to the other three (as previously mentioned). In a short time, they can both defend and attack, so defending their life.

p. 85 §3 When armed with a dagger, it should be positioned in such a way that it can dominate the enemy's sword; and with the left hand, be ready for defence.

1 *"4. When the body is in one and a half steps, forming an equilateral triangle."* – p. 17 §5.

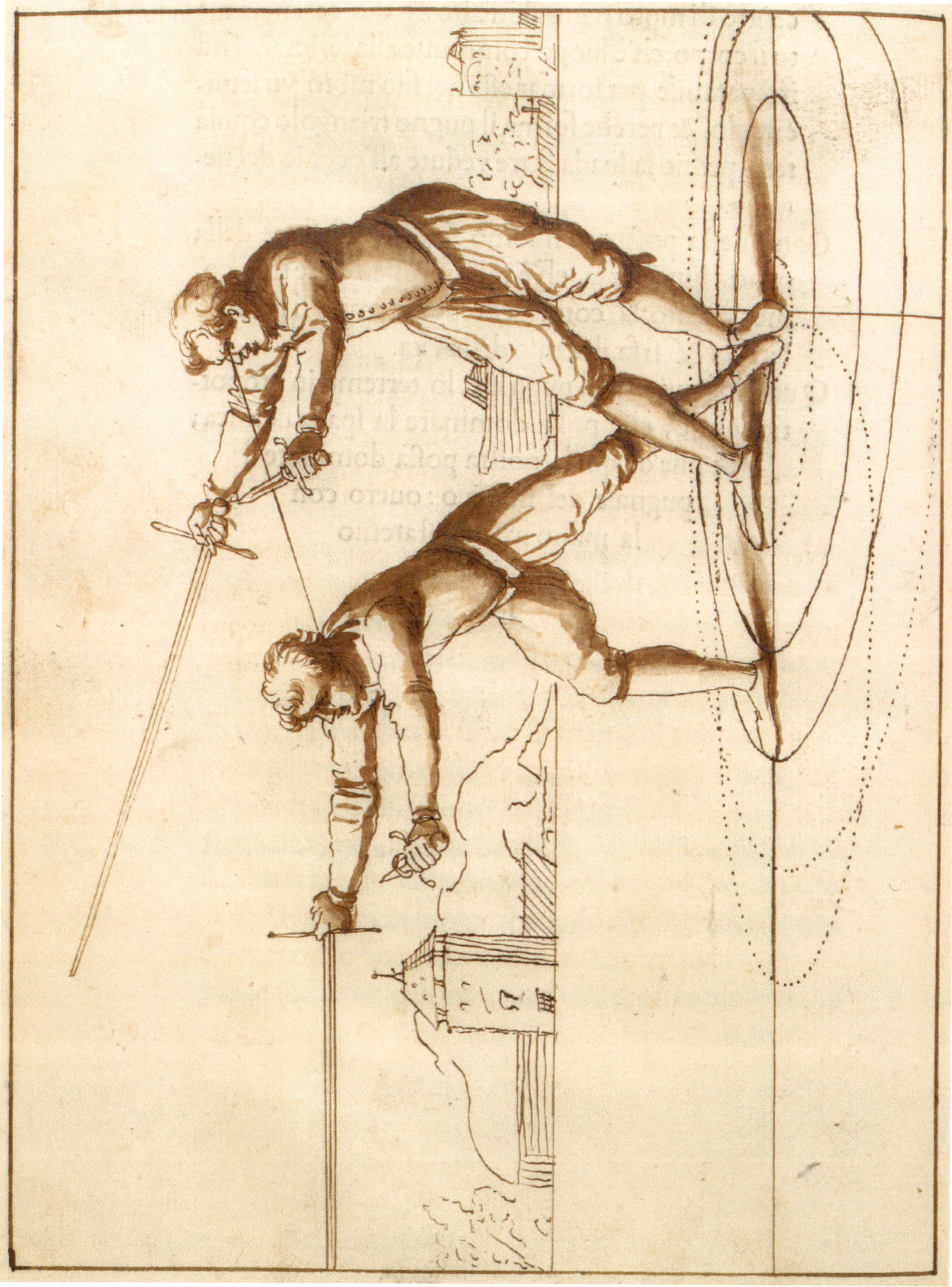

p. 87 §1 We should note, however, that while in this posture, we can transition into the other three postures, and likewise move from one to another as needed, simply by moving one leg and keeping the other still. For example, if the enemy's sword is aimed at our right side, according to the principle, once we've moved the point at which the enemy's sword is aimed, we'll also move the action by stepping with the left foot into the second circle,[1] which is done in a sideways motion. By shifting the body's weight onto the left leg, we form the second guard.

p. 87 §2 Since the right foot must move towards the attack, it's relieved of some of its weight in this guard. The sword naturally aligns itself for a thrust; the dagger, still dominating the enemy's sword, remains ready for defence. With the same motion and length of the line, we strike, as illustrated in the following figure:

1 *"[The second circle of motion is when] he stabilises the left foot and describes another circumference with the mobile one: in this, we enter by sideways movement to exit with thrusts."* – p. 19 §2.

p. 89 §1 Yet we have our body divided, as has been said, into two equal parts by the diameter of the circle (i.e., the right side and the left side), for as Boethius says in the book of Divisions,[1] any division must be of two members. If the enemy places their sword straight at our left point, we'll move that point out of the enemy's sword's reach with the right foot in a sideways motion within the second circle,[2] which we perform in motion, both with the feet and with the body in profile, forming the third guard, holding the right arm in a thrusting stance and the left arm well extended. This will have to shut out the enemy's sword, as seen in this figure:

1 Boethius was a C6th Italian philosopher. See *The Cambridge Translations of Medieval Philosophical Texts || BOETHIUS: On Division*, Cambridge University Press, #1, pages 11-38, 1989 Feb 24. Kretzmann, Norman; Stump, Eleonore, doi: 10.1017/CBO9781139171557

2 *"[The second circle of motion is when] he stabilises the left foot and describes another circumference with the mobile one: in this, we enter by sideways movement to exit with thrusts."* – p. 19 §2.

p. 91 §1 If you happen to be in any of these three guards and the enemy places their sword angled in presence, then with just a motion of the right foot, one will enter the fourth guard from those positions. (Although the description made below is self-contained, without the person being placed in any of those guards.)

p. 91 §2 The right foot is placed next to the left in the first circle with the body balanced, and by moving the right foot backward in a straight motion along the diameter of the circle formed on the ground, arriving thus measured in the third circumference of the third circle with the right foot,[1] the fourth guard is formed, with all the weight of the body on the right leg.

p. 91 §3 In moving the right foot back, a semicircle is made with the sword and it's placed sideways under the left arm (which is held outstretched and ready for defence). In that motion, which is made with the right, a foot of ground is gained with the left, to deceive the enemy. The body is kept on the right leg, to distance it from the enemy, and although the attack is made with the right because the body is what moves first towards the strike, it matters nonetheless that the weight is on that leg. This posture follows below:

1 *"Similarly, he forms the third [circle] while in the fourth stance [legs forming an equilateral triangle], which we enter when making the strike."* – p. 19 §2.

I'm not convinced that this is depicted in this illustration, which seems to lack the advance of the left foot, or has the foot too close together when the passing step is made.

For more information see p. 123 §1.

p. 108 §5 *So ends the Theory*[1]

1 Ghislieri's text has a 16 page section here, an extended analogy between fencing and siegecraft. To improve the flow of the book, that section has been moved to the end of the fencing content, after the Treatise Against a Left-hander. It's worth reading, but disrupts the flow when trying to understand the rapier and dagger positions and their use.

p. 109 §1

ON PRACTICE

p. 109 §2 *Our practice is reduced to a few things — for with our actions we deny the various principles behind all the attacks that might be used against us.*

p. 109 §3 As soon as the enemy is visible from afar, let the cloak fall from the right shoulder. We push the dagger from behind with the left hand, forward to the right side.[1] Then, first draw the sword from its sheath, and then the dagger. With that clench your fists, bare your teeth, and widen your eyes to show courage on your face.

p. 109 §4 With the weapons held low and extended away from the body, approach with natural steps. During the approach, carefully observe the enemy to identify which foot and which weapon they're presenting. If the enemy holds the dagger forward, it will be for defence; if it's the sword, it will be for offence. In any case, we'll always be the one who will attack and dictate movement.

p. 109 §5 Assuming the enemy is in a stance with their sword actually positioned in
p. 110 §1 the straight line, but controllable by our dagger, such as in the guards of 3rd or 4th, we'll present ourselves face-on and as we enter closely, adopt the first position. By adhering to the aforementioned principles, we can deliver a quick strike while at the same time beating the enemy's sword, diverting it from its line, aiding its descent.

p. 110 §2 For if we consider the four causes previously mentioned: the first being the posture of the body, we can be certain that the body in motion can move directly. The proportion of the sword, the formal cause, indicates to us the strike in a straight line. The material cause reveals that any other type of motion the enemy might make, whether with the body or the arm, whether pushing, withdrawing, carrying, or turning, will take more time than the motion indicated by the posture. The final cause, which is the endpoint of such a line, is now disrupted by the dagger as the beat diverts it from its course.

1 There are several methods of wearing a sword and dagger depicted in contemporary art. Some have the dagger on the right side of the belt, others behind the back. Perhaps the underlying principle is to have the left hand hold the sword belt still, out of the way of drawing the sword.

p. 110 §3 Similarly, so that one does not become confused and make errors, we should
p. 111 §1 understand that all the principles we've discussed apply as much to the play of sword and dagger as to the sword alone. However, there's a difference: when we have the dagger and are fighting against straight lines that are present but beneath our dagger, there's no need to carry away the points, which are the final causes of these lines. This is because by beating with the dagger, we negate the effect of the enemy's thrust.

p. 111 §2 But when we fight against the geometry of those with the sword alone, which lacks such defence, we'll always strike while moving the left foot, since we need to displace the point with the motion of the body due to the absence of the dagger. Additionally, we'll use the left hand to close off the return of the enemy's sword, as demonstrated by the two small figures in the first demonstration that follows below.

p. 111 §3 And against high straight lines in presence (both when accompanied by a weapon and with the sword alone), we'll act in the same manner.

p. 111 §4 When fighting against angled lines, we'll follow the same order. The only difference is that when we strike from necessity with the sword alone (that is, in the timing when the enemy is striking at us), we'll do so with a straight line and our body profiled.

p. 111 §5 Returning to the explanation of the Practice, I say that if the enemy is in 3rd
p. 112 §1 or 4th, whether with legs apart or close together, we'll present ourselves facing them, placing ourselves in the first guard with the left foot on the diameter of the circle. As soon as our line of sight, passing through the strong of our dagger, reaches the weak of the enemy's sword, we'll strike with determination, *ahead-of-that-timing*, targeting their right side. This is illustrated by the following figure, where, as in all other figures, it shows the guard with its strike made intentionally before the enemy's move. There's also a line drawn from the fist of our stance, indicating the strike *within-that-timing* made from necessity; that is, the method of striking when the enemy strikes. Additionally, the line originating from the enemy's hand indicates the initial position of the enemy's sword.

p. 114 §1 I won't fail to mention that if our enemy is in a stance with the straight line in presence, but well extended (as many do to keep the enemy at a distance), then we'll parry that sword with a beat of our dagger, describing an outward semicircle with it, so as to draw a greater distance from our body in the strike.

p. 114 §2 Whenever our opponent is positioned in the guards of 1st or 2nd (or in the facing guard, or entering guard), in such a way that their sword is in a straight line but high, we'll operate against them in this way: we'll present ourselves in profile in the first circle, and placing the right foot on the diameter of the circle, we'll choose to oppose the straight point. With the left foot, we'll move in a sideways motion into the second circle (that we form by our movement). Thus, we'll strike from the second guard, whether *within-that-timing* or *ahead-of-that-timing*, by beating with our dagger.

p. 114 §3 Our opponent could do three things:

p. 115 §1
1. First, as we present ourselves, they could finish moving that line and strike, but we'll strike with premeditated timing.
2. Second, he could strike us in the motion we make when we move with the point, which won't succeed due to science, because our motion is of brief duration so that we'll have acted before he sees it. Nevertheless, assuming this happens, we'll already have moved and have the hand and the dagger in place.
3. Third, they could withdraw when we parry facing on, but their withdrawal won't be of sufficient extent to counter our advance, since we're united; in this case, we'll strike with our usual thrown broach.

p. 115 §2 If the enemy changes guard before reaching close, we'll observe the manner in which they positions themself and proceed against them according to the rules mentioned.

p. 115 §3 I've stated this for clarity, although it wasn't necessary as all the figures presented below are founded on all the stated rules. The following figure demonstrates the effect of the second guard and the line that exits from the enemy's shoulder, signifying the angle of the opponent's sword, in which it's assumed it was positioned.

p. 117 §1 If the enemy were to be positioned in the same guards, but held their sword fixed at our left point, we'd immediately place our left foot on the diameter directly in front of the enemy's position; then, moving the right foot sideways into the second circle, we'll form the third guard and strike from high to low with a broaching, letting the left arm naturally defend as if it were a single piece, both with the hand and the dagger.

p. 117 §2 The enemy might change their sword position underneath when we go to strike, but by shifting the body weight to the left base and making a slight movement, we'll parry that sword and deliver the same strike.

p. 117 §3 I won't repeat all the reasons, since what's been said about the first guard applies to all straight lines. These figures will sufficiently demonstrate the effect, as this one below shows how to strike both *ahead-of-that-timing* and *within-that-timing*.

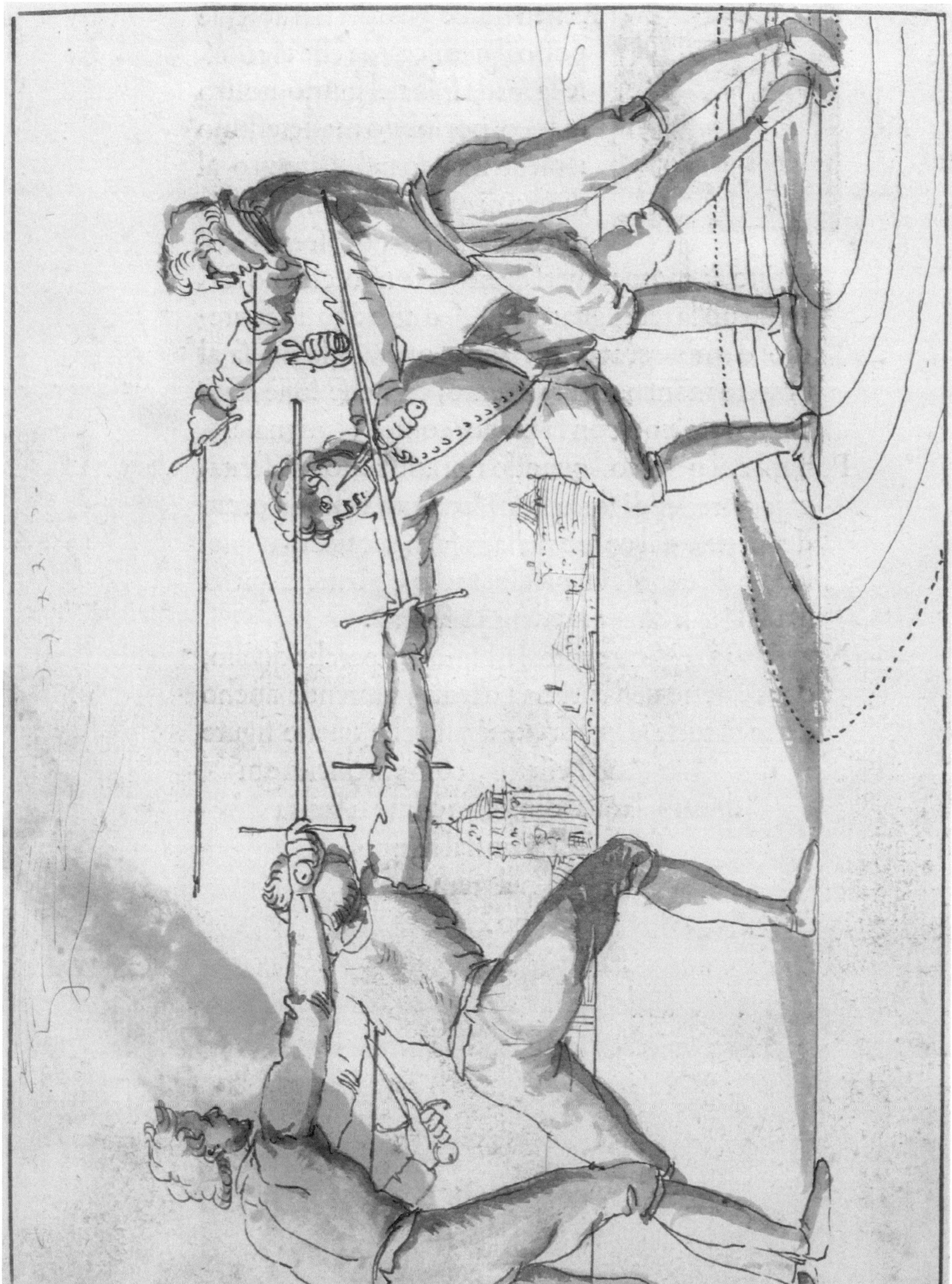

p. 119 §1 However, when the enemy has their sword in presence, but positioned angled underneath the centre of our body, then we should present ourselves in a unified stance with the body bending well into the first guard, opposing the dagger as a resisting force against the angled line to prevent it from rising. We'll then deliver our usual strike, chosen as it's the best one possible.

p. 119 §2 But be warned, for the main danger arises when we perform the beat while closing in. Therefore, with premeditated timing (since that angled line, in a short time, can't do anything other than rise into a straight line), we'll position our dagger above the imagined line where the sword is expected to strike. This imagined line is illustrated below in this figure.[1]

1 In the Bologna copy, the horizontal line between the fencers is labelled '*linea immaginata*' or 'imaginary line'

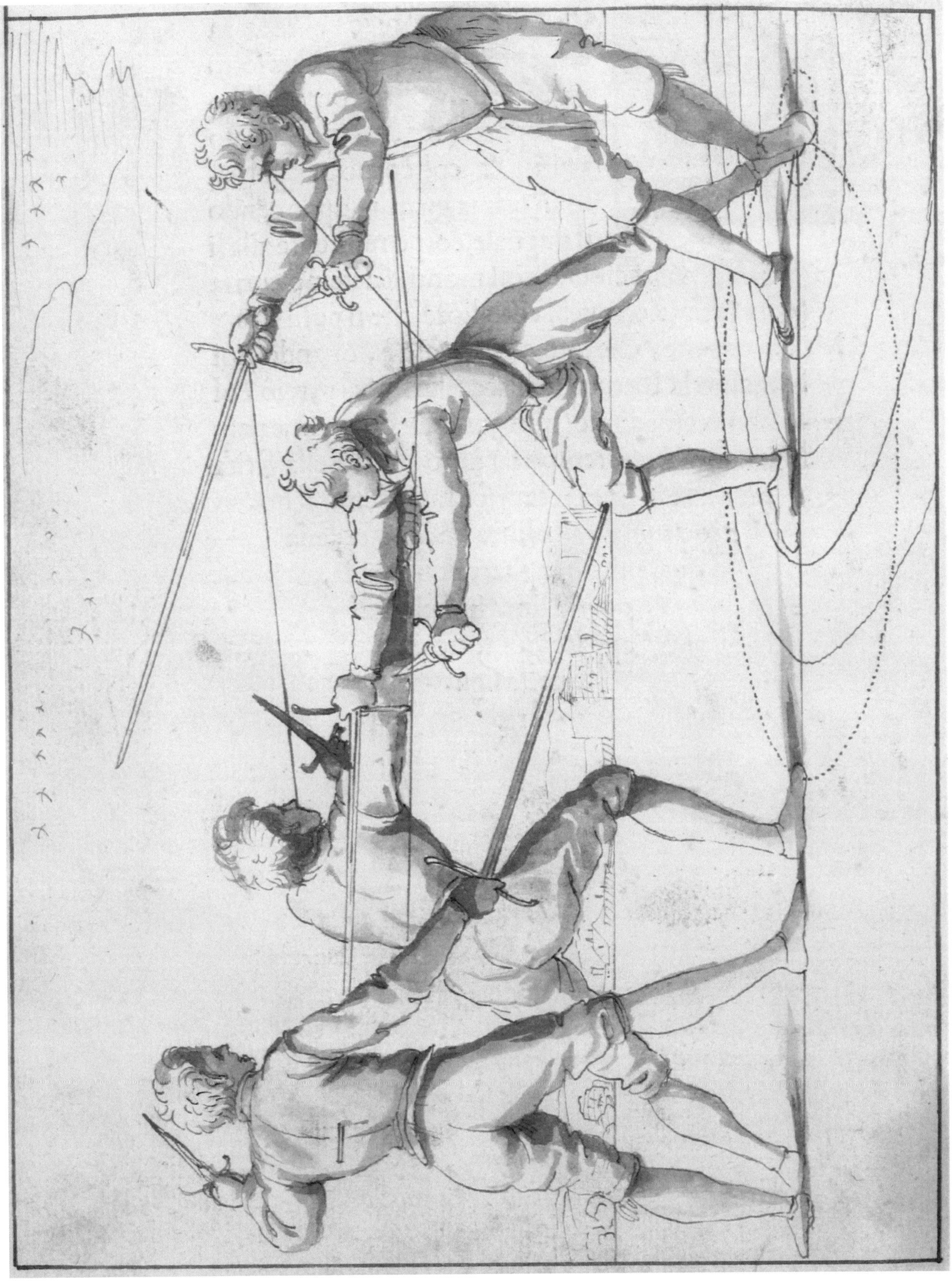

p. 121 §1 If the enemy holds their sword angled in presence in the second manner,[1] then immediately after we position ourselves on the diameter with the body facing them, we'll enter into the smaller portion of the circle of distance, which is marked with dots, just like the other small circle that we also form in this position. By placing our feet on the chord of the said larger circle, we'll form the first guard. In doing so, we gain the distance; in case the enemy strikes, we'll also strike, positioning our sword along the diagonal and our dagger on the imagined line where the angled sword is expected to come. This is well illustrated in the figure, which shows both the effect of our strike and the effect of the enemy's strike.

1 From the theory section, p. 71 §5:

"Secondly, the enemy's sword will be called angled from our body when both are in the diameter of the circle of distance, and the point of the enemy's sword is outside our body, either high, low, or in any kind of proportion; and this will be when the enemy holds the hilt of the sword at our right or left parts."

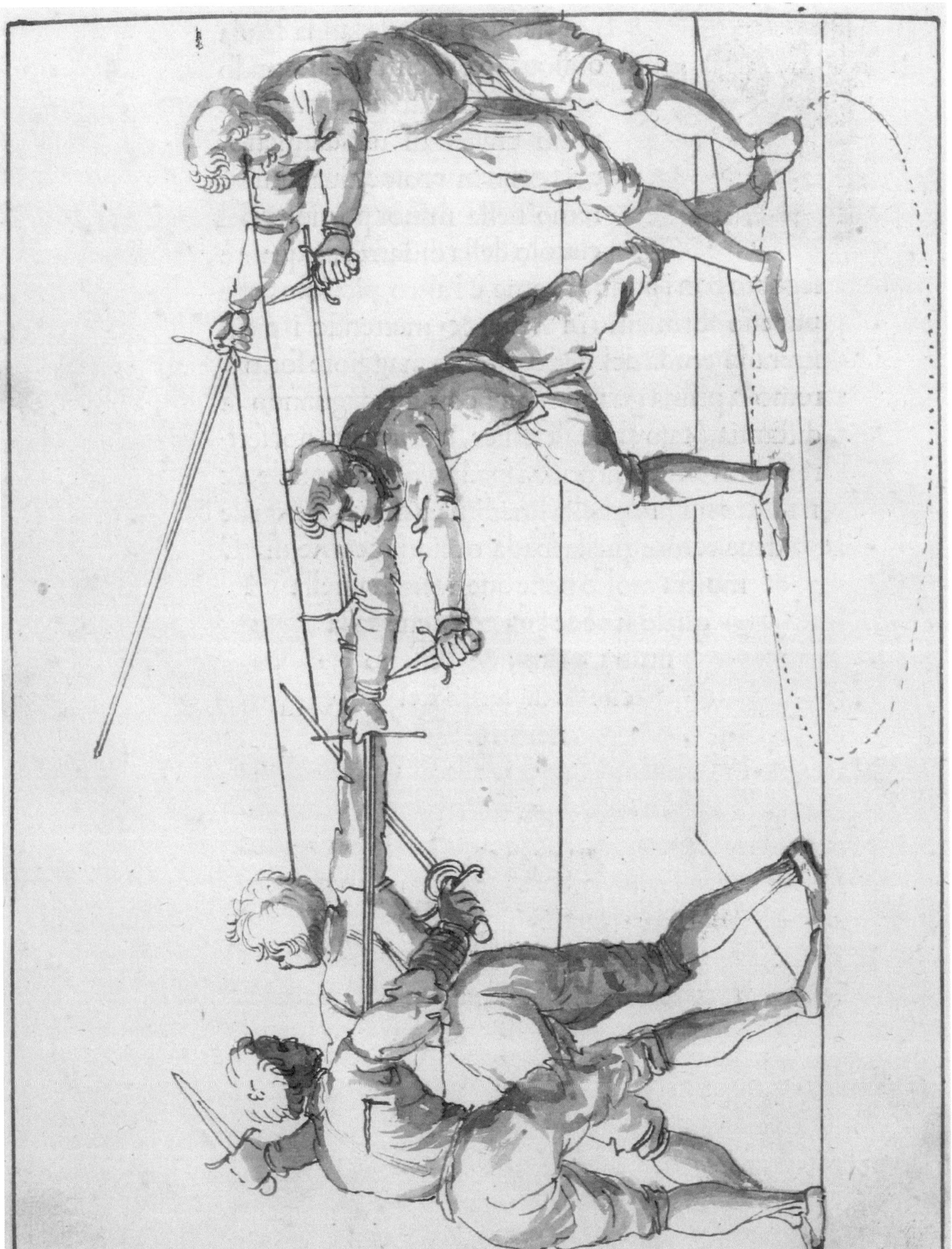

p. 123 §1 Our opponent may find themselves with the sword positioned slanted from our body in the third manner — that is (as previously described) at an angle.[1] We'll take note if the tip of their sword is directed towards our right or left sides, and as soon as we reach the distance, we'll adopt the fourth guard. We'll strike with the maximum extension of our body, moving it first, and then following up with the right foot, intersecting that line. We'll help it with a motion of the dagger at the circumference, assured that, by staying in this angle, should their sword become a straight line — we'll respond by shifting the weight of our body onto our left side, countering with a strike from the first guard. If, when we position ourselves, the angular line becomes straight, we'll return to the first circle in the first guard and proceed as we've described when discussing the first guard.

p. 123 §2 However, if their sword remains angled and directed towards our right side, placed such that we can see the body but not the tip of the sword, we'll present ourselves in the first guard, responding to it as if it were a straight line.

p. 123 §3 The figure below shows really well the effect of the enemy, and the effect of our own posture.

1 From the theory section, p. 72 §1:
"Thirdly, the enemy's sword is angled from our body when it forms an angle with the arm and its point is above the centre of our body."

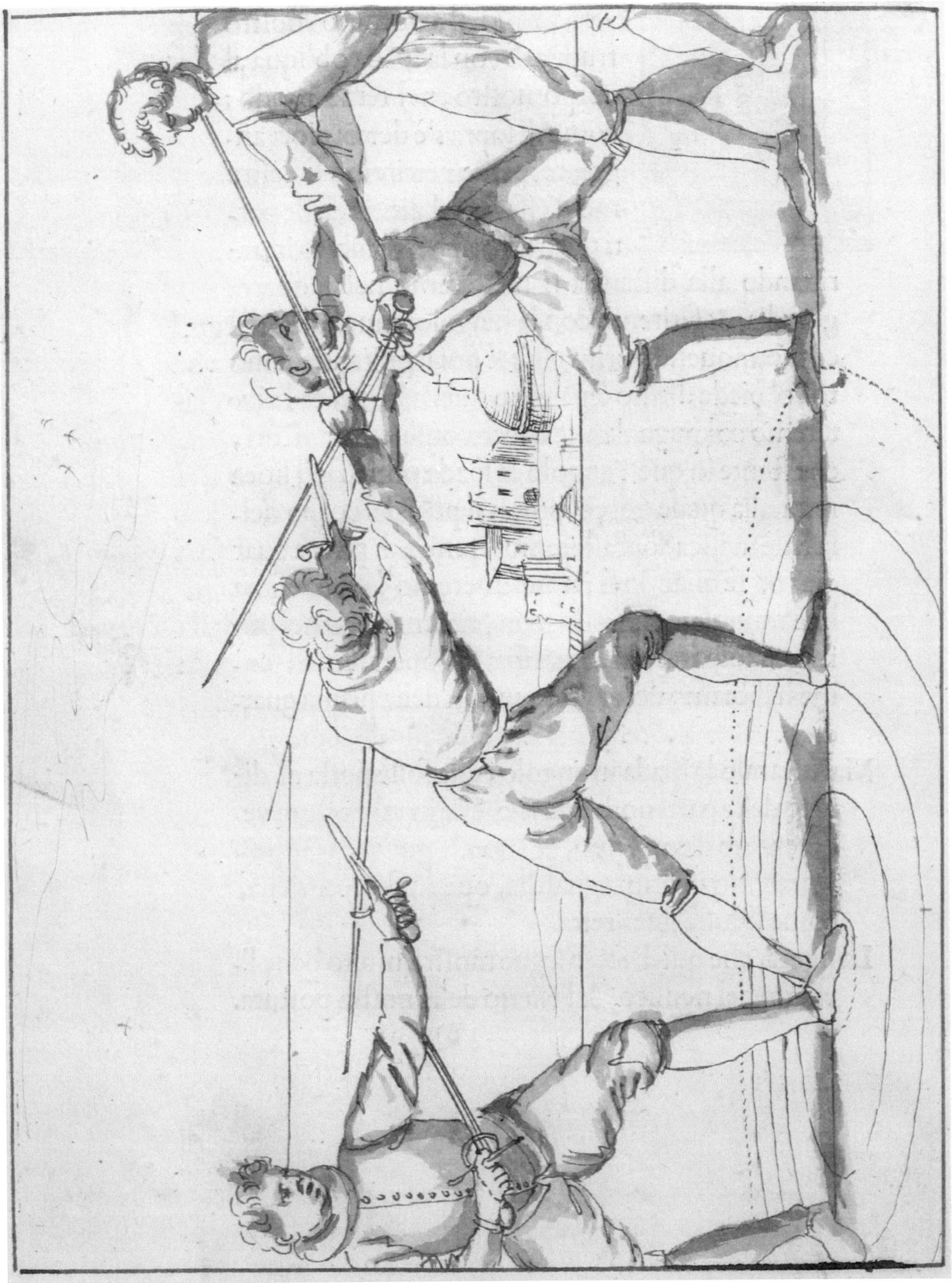

p. 125 §1 *Now that we've discussed the angled lines in presence, it remains for us to reach the conclusion of this treatise by addressing those lines that are out of presence, considering that the enemy may find themselves in such positions in one of two ways: either with the right foot forward or with the left.*

p. 125 §2 Since our goal is to strike our enemy it actually simplifies our action whenever they hold their sword back. We'll follow the rule of distancing ourselves from the enemy's sword by the same amount that it's distant from us. By doing so, we'll aim for the enemy's body, and by keeping our body foreshortened, we'll go out into the smaller portion of the circle. Standing on the chord of that circle, we'll strike *ahead-of-that-timing* and *within-that-timing*, in the same manner as shown in the following figure:

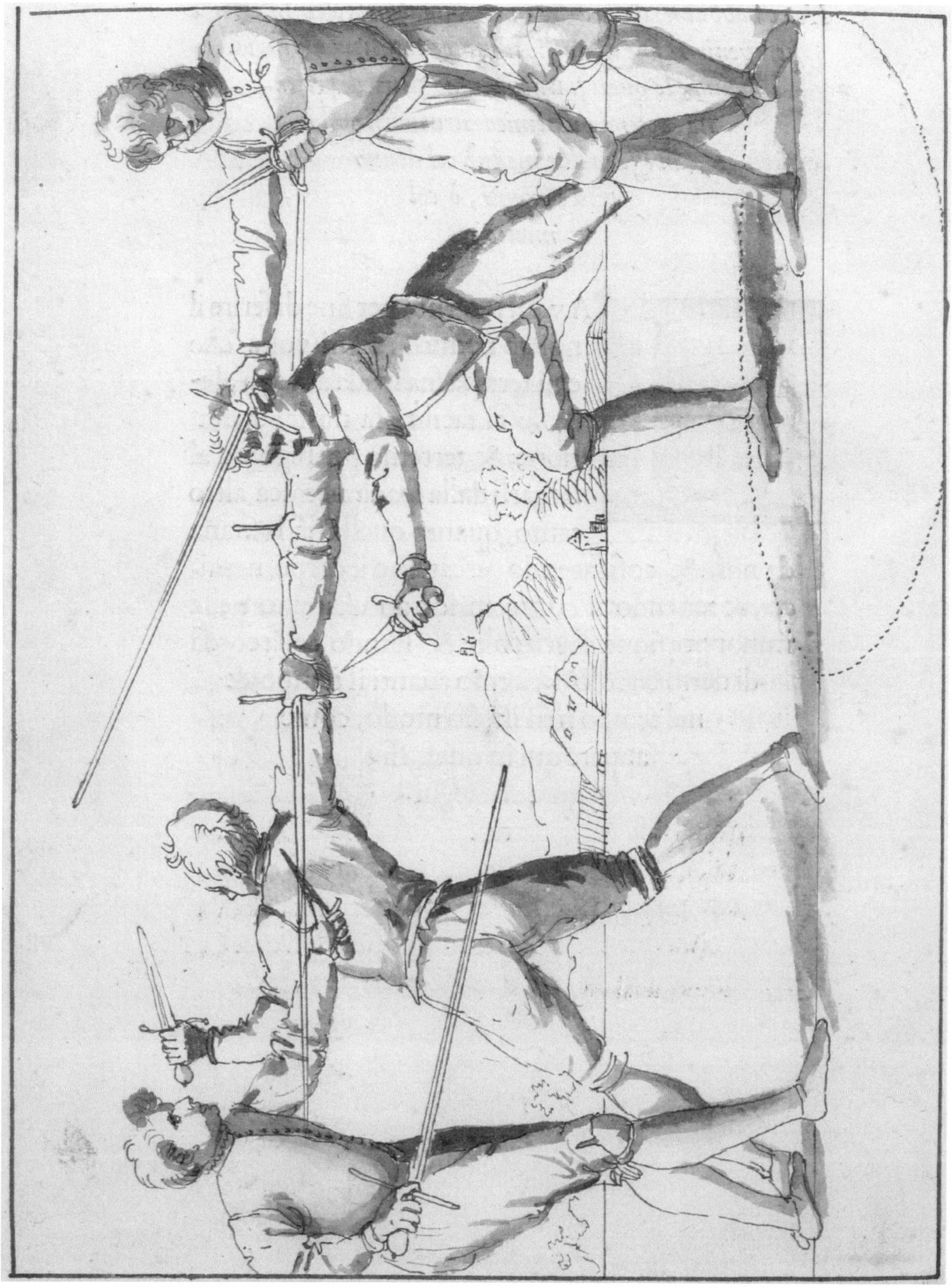

p. 127 §1 Finally, when we find the enemy positioned with their left foot forward, in this stance they only offer a single point of their surface. However, when they intend to strike, they'll have to move with the entire face of their body into the strike. This will be done with the greatest movement possible, which can be described as taking the time of a *Maxima*: that is, eight beats.[1] We'll therefore close the distance, positioning ourselves in the first guard with the smallest bodily proportions — that is, with our body turned closer to the profiled position than otherwise. Also, we'll hold the dagger high if the enemy is in the guard of the falcon.[2]

p. 127 §2 While we operate in this manner, we'll provoke a movement from the enemy, extend the point, gain the distance, bring our sword closer to the enemy, and with the dagger held ready (in a manner that it dominates the enemy's sword), we'll strike the enemy either *ahead-of-that-timing* or *within-that-timing*. To ensure this chapter is better understood, I've divided it and created two demonstrations. The first shows that when striking voluntarily, we no longer use our usual strike, but instead, we perform it with the body in profile, as seen in the following figure: and the strike is delivered to their left side.

1 See the fencing time theory, p. 51 §4.

2 A high Agrippan 1st position as illustrated.

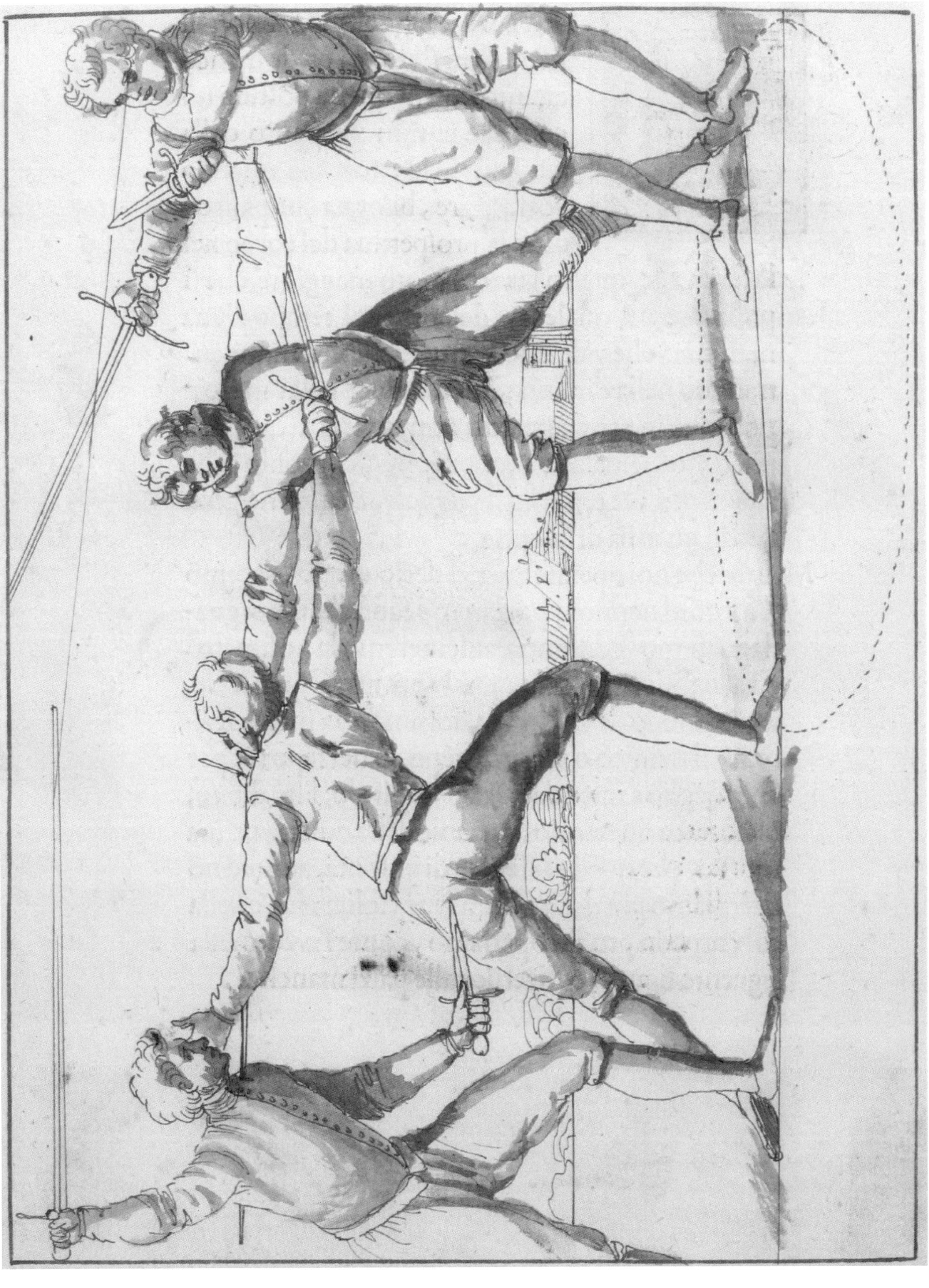

p. 129 §1 Whenever we provoke movement from those who are positioned with their left foot forward, they're compelled to adjust. This is because, by raising our point, we cover ourselves behind the line of their body. Thus, as they find no opening to strike, they're forced to move. This is the moment to strike them during that movement, whether it be with the right foot or the left.

p. 129 §2 But if they attempt to strike, we'll use our usual parry and strike to hit them on their right side, which is the side advancing with their strike, as illustrated in the following figure.

p. 131 §1 All these things we've discussed must be put to the test, according to the need. The enemy could (for example) position themselves in a guard of entering,[1] and then lower that line and, furthermore, free it and place it into an angled line. Likewise, they could shift from one position to another. However, by following the aforementioned rules, we can quite easily transition from one guard to another.

p. 131 §2 These considerations clearly show that putting oneself into a guard is very harmful. Consequently, in all our actions, we must never reveal our intentions. Instead, with full determination, we should position the dagger along the imagined straight line and strike with a single blow: either with a thrust (which should be a powerful lunge), or with a downward cut (using all our strength and reach, so forcing the enemy to parry). Immediately afterward, we'll strike again with a backhanded cut across, supported by the force of the dagger, and then return to the initial strike. This backhanded cut should be executed from the first opening, while maintaining the strong stance created by the initial strike. However, if the enemy retreats from our first strike, we'll deliver the backhand cut while advancing with the left foot.

p. 132 §1 This approach should be carried out with the understanding that reaching the distance and striking should be simultaneous actions.

p. 132 §2 By re-engaging resolutely and swiftly, we keep the enemy occupied with their defence, causing them to lose their strategy and spirit, and consequently their strength. It becomes clear that the enemy will either put themselves on the defensive, or attempt some form of involuntary strike. These strikes, made out of necessity because we've forced them to act against their will, can easily be resisted — especially since they're weak and imperfect.

1 Dall'Agocchie *Dell Arte di Scrimia* (1572) describes *guardia d'entrare* with the fencer's sword extended at the opponent's face, and the hand pronated so that the long edge faces outwards to the right. In Marozzo (1568 edition), Giovanni Battista Fontana's engravings show the hand positioned in 3rd or 4th.

p. 133 §1

CONSIDERATIONS[1] FOR SWORD & DAGGER

p. 133 §2 *When we position ourselves in our first guard, the opponent may attempt to counter this stance in five different ways. However, all of their efforts will be in vain if we act according to the principles and science of the art.*

p. 133 §3

1. The first way, the opponent will attempt to close the distance quickly and launch a strike around at our sword, aiming to strike with their dagger. The remedy to this is to strike the opponent in motion before they reach the distance.

   p. 133 §4 Furthermore, with every little movement raising our sword, we place it in the position for a straight-line strike. Additionally, we step forward with our left foot and strike with the dagger.

p. 133 §5

2. In the second way, the student will try the opposite of going for the
   p. 134 §1 sword: that is, delivering a backhand thrust. To counter this, we make the centre ourselves by stepping forward with our right foot, preventing them from closing in. In their movements, we have the opportunity to strike. If the opponent manages to execute the backhand thrust, we'll thrust our sword in a straight line with our body in profile, thereby blocking across their line.

p. 134 §2

3. In the third way, with their sword shrewdly lowered,[2] the opponent might attempt a '4^{th}er'.[3] In response, we'll move shrewdly on above and we'll block their sword from rising with the dagger.

p. 134 §3

4. In the fourth way, the opponent might raise their sword midway, not to strike, but hunting for a backhand thrust. The remedy to this is the same as previously mentioned.

p. 134 §4

5. In the fifth way, the opponent might position their sword's point against our hilt to prevent it from lowering. We counter this by striking as soon as their sword exposes their body. However, if their sword is already on our hilt, we feign a slight retreat by moving our sword's hilt; should they then attempt to adjust their sword, we strike them.

---

1 *avertimenti* – both warning or admonishment, and a prompt or cue.

2 *giostrando* – the same root as jousting, and can also mean to skillfully manoeuvre. Perhaps *jostling* gives the gist.

3 *inquartata* – 'en-fourth-ed', referring to Agrippa's guard of 4th.

p. 134 §5
- If one misjudges or becomes disconcerted while positioning oneself, we should retreat while executing a circular backhand cut that cracks the ground. After regaining composure, we should resume our guard.

p. 134 §6
- When the sword feels impeded, one should yield to that force and strike in the direction that the force dictates, for this is the best way of using arms.

p. 134 §7
- If the opponent counter-positions, we must also counter-position ourselves, for the one who goes second has the advantage.

p. 135 §1
- When you can strike with just four fingers of the sword, there's no need to try to overwhelm. The first strike should be delivered while retreating, especially when the opponent has a dagger.

p. 135 §2
- In all strikes, the body should extend fully first, followed by foot movement, ensuring that the point is quick to arrive at the strike.

p. 135 §3
- Every time their dagger parries our strikes, redo the blow; for it has been aided by the dagger's beat. That is, we won't force our arm from its natural position.

p. 135 §4
- Whenever we assume our stance and the opponent retreats (which is more often than not), we lower our weapons and resume the fight in our usual manner. If the opponent seems to hold their guard, we continue to restrict them as much as possible, forcing them into necessity.

p. 136 §1

CONSIDERATIONS FOR THE SWORD ALONE

p. 136 §2

- Every time the sword is engaged it can be freed by stepping the left foot back — and if the opponent strikes, we shall strike at the same time.

p. 136 §3

- If we come to the middle of the sword by making some strike, we shall free our own sword. If it's held on the inside, we shall raise it up, planting the strike, and we'll grasp the opponent's sword by its furniture with the left hand. But if it's held on the outside, we'll free downwards, plunging it into the angle formed by the two swords while our left arm remains across for defence, and we'll thrust with the sword, raising the hilt upwards.

p. 137 §1 TREATISE ON SWORD, AND CAPE.

p. 137 §2 When we use the cape for defence, always remember that it's different from a dagger. It can be cut and pierced, unlike a dagger. Therefore, we'll never parry with the cape in the way we'd with a dagger. Just as stabs with the dagger aren't parried, so with the cape we won't be able to resist them. Just as we try to keep the dagger free from being attacked, the cape mustn't be kept subject by our enemy.

p. 137 §3 All the same reasoning and methods explained before, for the sword alone
p. 138 §1 and the sword accompanied by the dagger, also serve us with the sword and
cape. There's no use for it other than impeding our enemy: that is, so it isn't as easy [for them] to strike, and to cover the lower parts.

p. 138 §2 Just as we use the hand to close off the return of our opponent's sword when we strike with the point, so the cape works better to achieve that end.

p. 138 §3 A person can harm and hinder the opponent in two ways by using the cape: one is by throwing it in their face, and the other is by throwing it over their sword, particularly on its weaker third. The latter is easier than the former. Therefore, the cape should be grasped with an open hand, using only the thumb, and then gathered into the fist with a semicircular motion from the outside, so that they can be used together as desired and also easily discarded to engage in grappling.

p. 138 §4 From this, it can be understood that when drawing the sword and wrapping the cape around the fist, the same order should be followed as is done with the sword and dagger. This must be done without revealing our intentions to the enemy — which we'll achieve by advancing with the sword held low and the cape extended away from the body. We'll engage with caution, observing the actions of the enemy, who'll either be static or in motion.

p. 138 §5 If the opponent is found to be static, they'll necessarily be in one of the
p. 139 §1 positions we've already described — so we'll use one of the offensives against
them that we've previously explained.

p. 139 §2 Should that same opponent hold their sword forward, we'll combat it using the methods already mentioned. However, if they hold the cape forward, we'll always move naturally to our right, striking with cuts and thrusts at the hand holding the cape while distancing ourselves from the opponent's sword.

p. 139 §3 If, however, we happen to find our opponent in motion, we'll respond according to the type of attack they make. Whether it's a cut or a thrust, we'll execute the same attacks and defences as if we didn't have the cape, except for parrying. In that case, we won't parry with the sword alone, but rather with the sword accompanied by the cape, and in a subsequent timing attack the right side of our opponent.[1]

p. 139 §4 Should we wish to wait for the enemy, we'll take note of which foot they have forward and present ourselves with the opposite one. For example, if they have their right foot forward, we'll position ourselves with our left foot forward, and if they have their left foot forward, we'll do the same with our right foot.

p. 139 §5 Pay attention that when the opponent has their right foot forward and we
p. 140 §1 oppose them with our left, this will be with the intention of letting all attacks pass by and striking during their descent with the support of the cape.

p. 140 §2 If the opponent stands with their left foot forward and we with our right, immediately opposing with left, we will then parry, striking at the legs or at the right arm or with a backhand thrust

p. 140 §3 In this case, we'll be able to parry cuts with the cape alone because, with the opponent's advance and our movement, we'll parry into the first third of their sword before it can descend into full force.

p. 140 §4 It can happen that we have to make up for an opponent's blow with our sword so that we can ensure our safety on such an occasion from such strikes. For if, by chance, we were to strike *within-that-timing*, it'd most often be aimed at a place where the opponent's cape will entangle and obstruct our sword, preventing us from recovering it in time. Thus, securing ourselves from the enemy's strike, we'll strike to the opponent's right side in a subsequent timing.[2]

p. 140 §5 To anticipate the opponent, we'll position ourselves in this guard, as seen
p. 141 §1 below: with the left foot forward in a half-step, the weight placed on the right leg, and the sword across. The tip of the sword should extend beyond our head, with the hilt level with the left shoulder, and the left hand, open and wrapped in the cape, placed on the strong first part of the sword.

1 This line is hard to parse – *"& di secondo tempo feriremo alla parti contrarie dirette"*. It could mean repeating an action, or doing an action in the next moment. It could also mean attacking directly, or with a forehand cut.

2 As per p. 139 §3.

p. 141 §2 In this posture, the enemy's sword will always be dominated inward, and as he delivers any forceful strike — whether a cut or a thrust — you'll only need to extend the left foot in a forced step. By extending the line, you'll strike, aiding the sword with the strength of the left hand.

p. 141 §3 This is demonstrated by the following figure [on the next page]:

p. 143 §1 Should it happen that the opponent passes to the right side, the responder, making use of the point in perspective,[1] must strike with a broach,[2] letting the cape close off the enemy's sword's return.

p. 143 §2 If, while we're in this posture, the enemy does not resolve to strike but instead, also putting themselves in guard, waits for us to strike first, with their sword in presence and their right leg forward, we'll press them with the same posture, always striking with the same attack.

p. 143 §3 However, if the opponent has their left foot forward and attempts to obscure our sword with their cape, we'll then return the left foot to the centre, exiting to the first guard among the four we've proposed, and strike that hand with a thrust or cut. Alternatively, we can move above, passing with the point to strike their face.

1 That is, turning face-on to the opponent while thrusting.

2 *Imbroccata.*

p. 144 §1

TREATISE ON THE BUCKLER

p. 144 §2 Many have used the buckler in the same way as one would use a dagger, and with it, they've parried, accompanying it with the sword. However, since it's difficult for one to demonstrate any skill by means of an instrument without first understanding what that instrument is, we state that, for those who wish to produce an effect with the proper instrument for that purpose, it's necessary, in order to avoid error, that they know the qualities and properties of the instrument itself with which they intend to achieve that effect. Therefore, having to discuss the buckler, an instrument (as I said) used for defence, we shall now begin our discussion of the principles.

p. 144 §3 The buckler is a spherical, dense, and opaque object. In order for it to properly
p. 145 §1 serve its function, its outer curved surface should have a 1' diameter. This is because, generally, men are about 1½' wide.[1] Since the buckler's purpose is to resist the sword, so that it can't wound our body, and since a sword can create many different and infinite lines, which can be likened to the rays of the Sun, a spherical and luminous object, we must understand that there are two types of opaque bodies: dense and thick, such as the earth, and translucent; like air, water, and fire.

p. 145 §2 Hence perspective, as outlined in the 24th proposition of the 1st part on common perspective,[2] it's thought that a spherical, dense, and opaque body (consequently, one that casts a shadow), can be considered in three different ways in relation to another spherical, luminous, and radiant body.

p. 145 §3
1. In the first way, it's considered when the body casting a shadow is larger than the luminous body. In this case, it's argued and demonstrated that the said shadow-casting body will cause and project a shadow which becomes wider the farther it extends. An example of this can be seen in the following diagram, where the luminous body is represented by **A**, the shadow-casting body by **B**, and the shadow itself by **C**:

---

1 I think that Ghislieri's intention here is to refer the buckler size to human proportions as per pp. 7-10, rather than a contemporary unit of measurement. That said, for more information on contemporary units see:
https://pieceofheaven1951.blogspot.com/2020/02/sometimes-massive-research-projects.html

2 This is referencing Euclid's *Optics* – *'When the eye approaches the sphere, the part seen will be less, but will seem to be more.'*

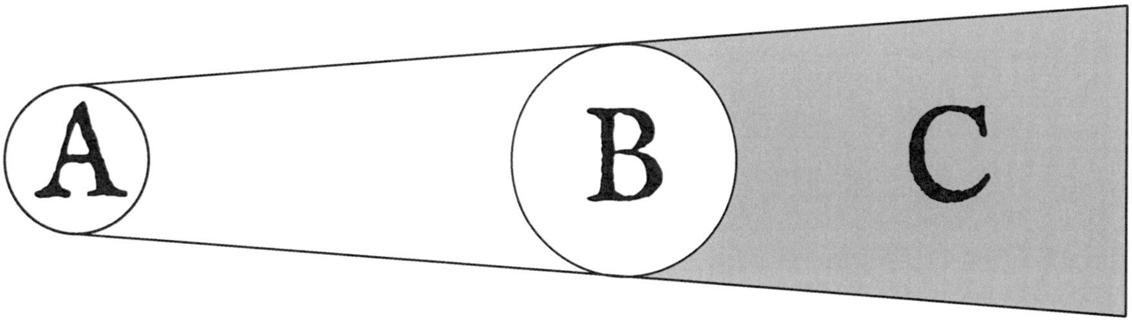

p. 146 §1 2. In the second case, we must consider that the shadow-casting spherical body is the same size as the luminous one. In this scenario, the shadow that is cast will always be of equal width to the diameter of the shadow-casting body, no matter how far the shadow extends in length. We can clearly observe this in the following figure:

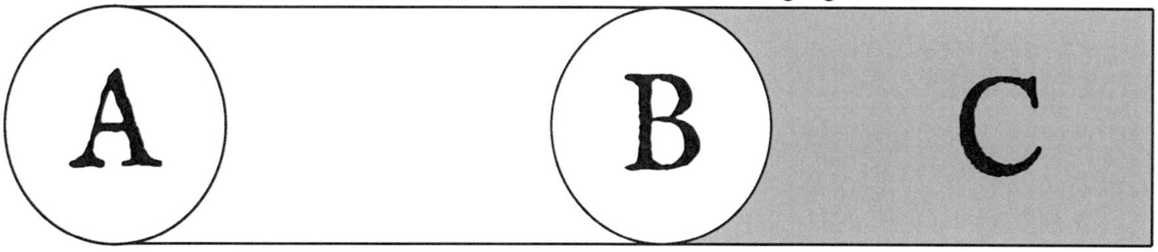

p. 146 §3 3. Lastly, in the third scenario, we can consider that the shadow-casting spherical body is smaller than the luminous one. In this case, the shadow will narrow as it extends in length, until it eventually comes to a point. This results in a pyramidal shape. The extreme tip of this shape is called a *Cone*. This phenomenon is illustrated in the following figure:

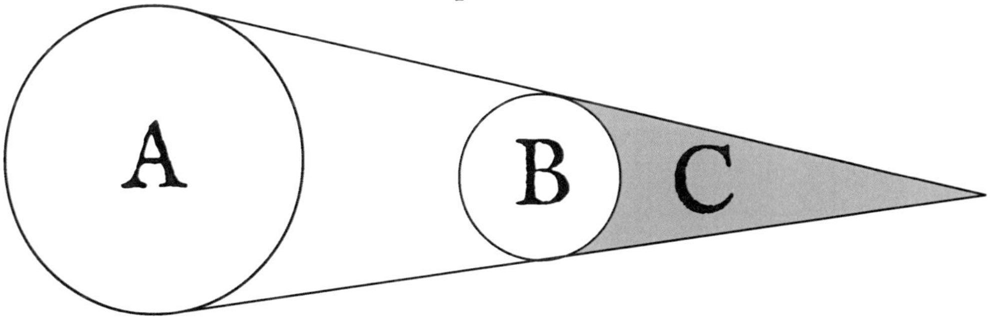

p. 146 §5 Of all these methods, the first is the one we'll use with the buckler. Since it's a
p. 147 §1 spherical shadow-casting body larger than the luminous spherical body — that is, larger than the fist from which, like the rays of the Sun, the proportions of the lines formed by a person wielding a sword emerge — we'll position the buckler in such a way that it casts a shadow under which we'll be covered and defended.

p. 147 §2 We'll nonetheless distinguish in this instance that the opponent will either have their sword in action along the straight line or in potential — that is, in an angled line. When it's held in a straight line, we'll consider this as a single ray of the Sun, and we'll oppose the buckler to it at that very point. But when the sword is held angled, away from our body, we'll oppose the buckler to the opponent's fist, as it's the source of the ray or the line; this way, we'll be covered by the shadow of the buckler's opaque body.

p. 147 §3 For just as many straight lines, all of which are coming from a centre, spread further apart from one another the further that they move from that centre, so by opposing the buckler to the sword, it will either strike the buckler or be deflected by it, causing the sword to be blocked from its centre and from our body.

p. 147 §4 Thus, the buckler will be held in hand with the thumb placed as support, so
p. 148 §1 that part of it does not give way inwardly; otherwise, our body would become exposed and vulnerable to injury. So, to ensure that it always functions like that spherical object, we'll hold it with the arm extended, so that the shadow it casts will be larger due to the distance between it and our body.

p. 148 §2 Therefore, with the buckler, as with all other sorts of defensive weapons, we'll act with composure, without giving the enemy any indication of our intent, while distinguishing whether the enemy is in motion or standing still.

p. 148 §3 If he's in motion, or if he throws either a cut or a thrust: to beat these cuts, we'll perform the same parries and strikes as previously described but using the buckler for parrying. We'll proceed with the weapons coordinated, following the centre of the opponent's hand, and stepping with the left foot to counter-position.

p. 148 §4 It goes without saying that we could strike *ahead-of-that-timing*, *after-that-timing*, or *within-that-timing* with the same methods as discussed in the treatise on sword and dagger.[1] However, we'll be mindful to always position the buckler near our sword hand, extending the arm and body well and shielding ourselves behind it.

p. 148 §5 If the opponent thrusts, we can execute the same strikes and parries that
p. 149 §1 have already been discussed. However, since objects in motion can easily be influenced by additional movement, we may use the buckler in defence in the same way we'd use a dagger.

1 See the theory section for that discussion of the timing of fencing actions, p. 52 §7.

p. 149 §2 Then, if the enemy is in a stationary position, they'll either hold the sword forward in guard or backward. If they hold it forward in guard, the sword will either be in a straight line or at an angle.

p. 149 §3 To all high lines, we'll oppose them with our buckler at its furthest reach and strike from below with our full reach, in a single beat.

p. 149 §4 For all straight lines in presence which are positioned directly toward the centre of our body (as when the opponent is in 3rd or 4th), the buckler will be positioned at its furthest reach and our sword will be directed upward into a thrust along the straight line.

p. 149 §5 To the contrary, if the opponent holds their sword in presence but at an angle from our body, the buckler will be positioned opposite the opponent's hand and we'll strike above in the closest available spot, using the same action.

p. 149 §6 In the event that the opponent holds their sword angularly, the same offence will be executed as if we were using a dagger, aiding the opponent's sword into its natural inclination.

p. 149 §7 Whenever we find the opponent with their sword positioned behind, we'll seek their body in the previously mentioned manner. We'll hold the buckler outside our body to intercept, as if it were a dagger, forcing the opponent's sword to pass through it.

p. 150 §1 When the opponent stops with their left foot forward and has placed their buckler in front, we'll approach in a straight, natural movement. Upon reaching the proper distance, with our body angled, we'll seek the opponent while distancing ourselves from their sword. By moving our sword in circular motions around their buckler, once we reach the level of our shoulder, we'll thrust our sword's point from above. If they raise their buckler, we'll thrust from below.

p. 150 §2 Additionally, we can position the point of our sword in the centre of the opponent's buckler. The opponent will likely lower their buckler to see our sword, and in that very moment, we'll thrust, directing the point toward their face.

p. 150 §3 For better understanding, I've included this figure here, to show the effect that the lines produce as they cross the buckler:

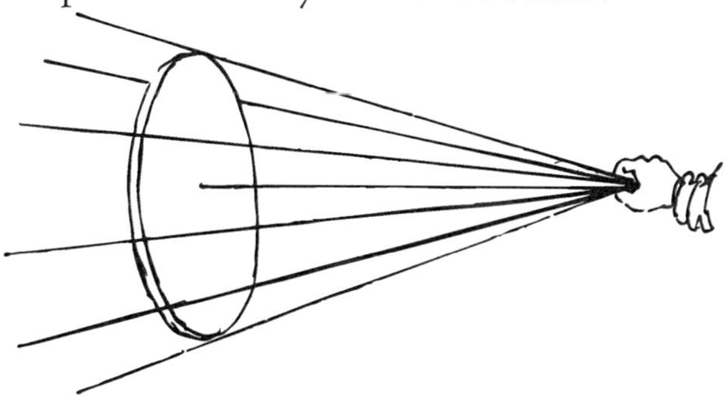

p. 151 §1

TREATISE AGAINST A LEFTY

p. 151 §2 All the reasoning mentioned for the single sword, sword and dagger, cape, and buckler apply equally to a Lefty[1] when fighting a Righty, and for a Righty against a Lefty. Defensive weapons are only useful for parrying strikes made from afar and for closing off lines of attack. This is because when a Righty faces a Lefty, both combatants have their swords aligned on the same side, and their bodies positioned such that all thrusts strike at angles. As a result, defensive weapons alone can't effectively parry these strikes, meaning the sword must be used for parrying. In these situations, the weak part of the Lefty's sword will naturally cross with the strong part of the Righty's sword.

p. 151 §3 Therefore, the Righty should hold their sword in a position that dominates the Lefty's sword inwardly, keeping their body in profile, and drive in the point to attack the nearest place with the front edge[2] of the sword.

p. 151 §4 The Righty will never strike the Lefty from the inside unless they're static — then they can be struck in that timing.

p. 152 §1 The Righty must distinguish whether the Lefty holds the sword in front or behind, and act against it in the discussed manner: that is, dominating the sword when it's positioned forward, forcing the Lefty to move, to strike in that timing. When it's behind, the combatant will ready their body to make the same strike, using defensive weapons to block the path of the Lefty's sword.

p. 152 §2 The Righty will let all cuts pass and strike *after-that-timing* of the slash, or aid it in its descent and strike a backhand. It's also possible to cover with a parry, and to make all the same strikes, *ahead-of-that-timing*, as we've discussed.[3]

---

1 The typesetter starts using capital letters for Mancino, but quickly changes to small capitals for both Mancino and Diritto. I assume that this is to clarify that it's a Righty and a Lefty, rather than a right part or a left part. It's neat.

2 This takes a bit of unpacking, and I took the *buono filo* to mean the 'true edge', the edge of the blade which normally faces away from the fencer. Ghislieri's advice to take the outside line, and attack with opposition in 3rd, pushing the Lefty inwards, is very effective fencing advice.

3 See the theory section for that discussion of the sequence of fencing actions, p. 52 §7.

p. 93 §1

AN EXTENSIVE ANALOGY OF FENCING AND SIEGE-CRAFT.[1]

Now, having described the theorems of this art, we're left to demonstrate the practice: but before we describe it, since this brief digression I'm about to make is so fitting to our purpose, even though it means entering into another subject, and since it will bring much pleasure to those in the profession, and furthermore, it will serve for greater understanding, I've resolved to briefly treat in general the attack and defence of fortresses, and only those that are pertinent to this purpose: and to then apply all of this to our particular skill, so as to learn to defend our body and to offend that of the enemy.

p. 93 §2 Since fortresses which were founded on naturally strong sites have proven weak through experience, man has ingeniously used art to choose the plain
p. 94 §1 for constructing a fortress that is less susceptible to the defects that've become evident due to new methods of offence. This has been done with the varied order of bastions,[2] cavaliers, platforms, flanks, curtains, shoulders, parapets, earthworks, casemates, countermines, moats, and ramparts; and with as many other parts and elements as are seen in the books of Teti, Tartaglia, and Castriotto, who've all written on this subject.[3]

1 This section was moved from the end of the Theory section, to improve the flow of the fencing content for the modern reader. I also added a title, to fit the format of other sections of the book.

2 *Bellivardi* – Bastion (or *Baluardo* – Bulwark), see "Some Architectural Drawings of Giovan Battista Aleotti", David R. Coffin, *Journal of the Society of Architectural Historians,* Vol. 21, No. 3 (Oct., 1962), pp. 116–128 (13 pages). doi: 10.2307/988055

3 Antónia Fialho Conde & M. Rosa Massa-Esteve (2018): "Teaching Engineers in the Seventeenth Century: European Influences in Portugal", *Engineering Studies*, doi: 10.1080/19378629.2018.1480627

Carlo Teti/Theti (Nola, 1529 - Padua, 1589) attended the university of Naples. During his career he worked as an engineer for the Habsburgs in Hungary, and in Italy for the Savoyards in Turin, the Medici in Florence, and the Este in Ferrara. He worked on the fortresses of Bergamo and Verona. He published in Venice in 1575 and Turin in 1584 and 1585.

Niccolò Tartaglia (Brescia, c. 1499 - Venice, 1557) was a Venetian engineer, mathematician and translator, who published six volumes between 1556 and 1560.

Giacomo "Castriotto" Fusti (Urbino, 1501/1510 - Calais, 1563) was an engineer who worked for the Duke of Urbino, the Habsburgs in Naples, the Papacy and the King of France. Three volumes of his works were published posthumously in Venice in 1564.

p. 94 §2 But nature, having formed the human body with such measured proportion, as mentioned, has likewise constituted within it a fortress composed of the varied order we've discussed, just as fortresses are constructed. It has two hands, like bastions, which defend the curtain wall — that is, the body. Likewise possesses other parts as will be seen in the following treatise.

p. 94 §3 For if a fortress were to be impregnable it would have to withstand all attacks upon it, and itself attack the enemy in a way that would not let them approach in any way. This hasn't be achieved so far, since the attack is made in a straight line. The path with which those inside may attack their enemy is the same by which those on the outside may expose theirs.

p. 94 §4 In consequence of this, the human fortress must possess these two conditions: firstly, to always keep its means of offence free and not subjected to the enemy's attacks; secondly, to not allow the enemy to come within its reach. Thus, we shall strive to keep the sword and dagger not subjected to the enemy's offence, but out of presence, as discussed regarding the first posture, as discovered by the Most Illustrious Sir Silvio Piccolomini.[1]

p. 95 §1 Just as a bastion, when deprived of its means of offence, can't support or defend another bastion, nor can it defend its curtain wall. Likewise, whenever one of the hands (which are like bastions of the human curtain wall) is so hindered that it can't execute its offensive role, the other will be left vulnerable to the enemy. We'll remedy this defect by keeping them free (as previously mentioned) and ready to perform their duties.

p. 95 §2 Not allowing the enemy to approach the moat is considered by all skilled soldiers to be the best defence. For, once the moat is lost, the fortress is said to be half conquered. Therefore, it's the opinion of skilled men that every effort should be made to defend the covered way, which, if well constructed, can easily be defended both from the walls with arquebus fire and by maintaining good soldiers in it who, at the right time and place, will sally out against the enemy.

1 See the positions in the Theory section, p. 83 §2.

p. 95 §3 From this principle of keeping the enemy at a distance, we must continuously
p. 96 §1 operate in such a way that our opponent never comes close to placing their foot on the circumference of the second circle, which we form while stationary (as previously mentioned), representing the edge of the moat of our fortress. This is achieved by delivering well-aimed thrusts from the walls — that is, from the posture without opening the compass — and with sorties, by opening our columns and entering the fourth circumference, which we form in motion, stepping with measured force (as mentioned), so that we can pre-empt the enemy with greater reach.

p. 96 §2 Sorties are conducted when it rains, or after keeping the enemy weary with continuous engagements in arms; and at supper time — whichever is best judged by those on the ground at the time.

p. 96 §3 Similarly, when the enemy finds themself exhausted from movement, we'll strike them by emerging in the fourth circumference, since they won't be able to respond with a counter-attack while they're in motion.

p. 96 §4 When the sortie is carried out, a selected group of well-armed soldiers is chosen and secretly led to the covered way. The enemy is then attacked suddenly,; and to ensure that these soldiers are secure during the retreat, the walls are armed with arquebusiers who are tasked with defending them.

p. 96 §5 In the same manner, we'll choose the best opportunity among our attacks and strike the enemy unexpectedly; since, seeing us so far from their body, they won't expect such a blow. To ensure our safe retreat, we'll strike them with a sideways backhand as we retreat to the covered way.

p. 96 §6 However, an attack that can't be carried out against fortresses during the day
p. 97 §1 can more easily be executed at night. By gaining the escarpment, one subsequently reaches the moat. Since a true attack is one where we're certain of not being counterattacked, ropes should be pulled and trenches dug. This way, we can approach the moat safely and under cover.

p. 97 §2 Thus, while we're attacking our opponent, we'll always make a semicircular motion with our dagger, which will defend our body.

p. 97 §3 The trench will dominate the inwards angle of the bastion so that it's not breached; it must also be brief, so that if the enemy discovers the intentions of their opponent, they don't have time to prepare a defence.

p. 97 §4 Similarly, if our enemy attempts to dominate our sword or dagger while covered (since the hands, which are our bastions, are mobile), we won't allow it. Instead, we'll continuously dominate the sword and dagger of our opponent from the inside, ensuring that our body is not breached — that is, wounded.

p. 98 §1 *Given that our opponent has dug the trenches and has approached the ditch, intending to use them to enter the fortress, it will be necessary to clear the way with an offensive strategy.*

p. 98 §2 Fortresses are attacked in three ways: either through hidden underground methods, such as tunnels and mines; or through open above-ground methods, such as battery, pickaxe, and shovel, and with explosive charges; or through the air, such as using raised platforms to prevent the defenders from appearing for the defence.

p. 98 §3 Similarly, our fortress can be tested in three ways: with low strikes, to the middle, and with overhead strikes.

p. 98 §4 When fortresses are attacked with tunnels and mines, the defender seeks to identify the location, which is done using various instruments. Because, as those below dig, these instruments provide indications of that activity. Once the defender is certain of the location, they either cut off that part, separating it from the main body of the fortress; or counter-tunnel to neutralise the mine, causing it to vent; or they attempt to destroy it with water or fire.

p. 99 §1 In the same manner, when we're attacked with low lines, such as when the enemy executes a '4ther',[1] we'll counter them by bending at the waist and not allowing them to penetrate the perimeter of our body.

p. 99 §2 Open ground assaults primarily utilise artillery. Artillery creates breaches and removes defences, thereby gaining the trench, and once the trench is secured, further attacks can be made with pickaxe, shovel, and explosive charges.

p. 99 §3 Therefore, artillery should be positioned as close as possible, if the location allows, to the place that needs to be targeted. For whenever a mass is propelled through the air by a violent force, the closer it is to the source of that force, the greater the effect it will have in striking a resistant body. This effect is crucial for achieving the objective in a short time, which is most valuable in attacking fortresses, and it also reduces expenses.

p. 99 §4 Thus, we should strike from as close as possible — that is, when the enemy is far enough away that we can be confident of reaching them by advancing with force. For if we begin the attack from too far away, we'll waste time, show cowardice, and, by exposing ourselves, give the enemy the opportunity to strike us. If we find ourselves somewhat distant, we'll advance first with the left foot and then deliver the strike with the right.

1 *inquartata* – "en-fourth-ed", referring to Agrippa's guard of 4th.

p. 100 §1 The greatest effect achieved by artillery in its strike is when the ball hits the flat surface of the wall at a right angle, as the diameter of the ball fully engages in the impact. When the ball strikes at an angle, part of the diameter makes contact while the other part goes astray.

p. 100 §2 Similarly, a strike should be executed at a right angle, meaning that the sword, in its strike, should form a right angle with the flat surface of the body. This way, the opposing sword becomes angled from our body. Additionally, when strikes are angled, even the slightest movement of the body can evade them and render them ineffective. We'll always strike at a right angle if we adhere to the previously mentioned rules.

p. 100 §3 The place where the artillery should be aimed is the face of one of the bastions, since removing the defences of the other bastion that defended that part and those above it allows for a secure advance for the assault.

p. 100 §4 Similarly, we'll target the enemy's bastion, which is the hand they present forward, and consequently, we'll keep our own hands away from harm.

p. 100 §5 Some people believe that the curtain wall should be targeted because it's
p. 101 §1 weaker than the flank and the face of the bastion, and due to its breadth and the space it has for retreat. This advantage allows defenders to stand above the attackers, which is not the case when the curtain wall is struck since there's no place for retreat except on the plain ground and the path is wider. However, they don't consider that this approach places one in the middle of five offensive actions: four from the flanks and one from the front.

p. 101 §2 Therefore, it's always better to attack the parts that are less defended. Consequently, we'll always strike at the part of the body that's most extended and closest to us, since it will be the weakest, just as in fortresses, and will lack defence.

p. 101 §3 We observe that the most defended part is the curtain wall, as the bastions are constructed to defend it. The part that forms an angle with the flank is the most protected. Thus, we'll always allow the enemy to target the point of our curtain wall, which is the right flank, so we can position ourselves in greater defence and compel the enemy to endure multiple attacks.

p. 101 §4 So, when the enemy aims to strike in this area with horizontal attacks such as thrusts, the left-hand bastion will deflect this line, and the right-hand will deliver a strike.

p. 101 §5 If someone attempts to strike the left-hand bastion, they'll be forced to pass their body's facing along the imagined straight line, where the right-hand bastion will intercept, providing an opportunity to strike them.

p. 102 §1 Similarly, if an attack were to be made on the right, we'd defend with it and attack with the left.

p. 102 §2 Aerial assaults occur when an external mound is built, allowing one to stand above those inside, who can't appear to defend themselves. This attack is significant because it targets the curtain wall and exposes the interior, making it the most formidable form of assault.

p. 102 §3 Thus, we'll always stand with our sword like knights against the enemy, attacking with thrown broaches,[1] going from high to low.

p. 102 §4 The defenders can also elevate themselves if possible, create counter-batteries, or construct traverses. They can take timely actions to obstruct and hinder the enemy's construction or even tunnel underneath with a mine.

p. 102 §5 Therefore, in response to the enemy raising their sword, we'll elevate our dagger to always stay above the opponent's sword.

p. 102 §6 Returning to the topic of battery attacks, after a breach is made, a skilled
p. 103 §1 soldier is sent to inspect it. They must assess whether the ascent is manageable, the pathway through the ditch is secure, and if there are any casemates in the flanks. Inside, they should observe if the enemy has retreated and the condition of the interior. Based on this soldier's report, the captain makes a decision.

p. 103 §2 In the same way, after we've inflicted a wound, we'll seek to understand the enemy's response. Given their potential anger from the strike, we must be vigilant and more prepared for a second strike than the first.

p. 103 §3 Retreats are executed in four ways, all achieving the same effect:

1. The first method involves digging a straight trench along the length of the breached wall, constructing casemates on the sides of the trench with timber and earth, and reinforcing the embankment with thick, pointed logs and gabion parapets.
2. The second method uses a crescent shape.
3. The third is an angular shape.
4. The fourth a pincer shape.

It's essential to leave as little ground as possible for the enemy, in order to prevent them from placing artillery or setting up positions above.

p. 103 §4 These retreats, when there's enough space to make them, are much better
p. 104 §1 than simply filling the breach. Working can only be done at night and sometimes poorly, making the defence weak.

1 *Imbroccata.*

p. 104 §2 Similarly, after receiving a wound, we'll retreat three steps while always keeping the body facing the enemy, to prevent them from striking us as we stand firm at a distance. Our retreat will form a crescent shape.

p. 104 §3 Thus, when the retreats are properly executed, they're challenging to overtake. Any proportion of line entering a circumference must always enter that circumference.

p. 104 §4 Therefore, we'll always maintain a retreating stance, with the body in a semicircle, forcing the enemy into the centre of our retreat. This retreat will consist of the sword and dagger in our guard. Any wound that enters the middle of our weapons will be blocked by the dagger, and the enemy will be struck by our sword.

p. 104 §5 When entering these retreats, we move with a French trench, which is
p. 105 §1 extremely difficult and strenuous. The earth must be thrown to the sides to cover, and fascines or planks are placed in front. The defender, by positioning artillery, causes significant damage and counters with various devices to push back the enemy.

p. 105 §2 Similarly, if our enemy, out of desperation, attempts to enter our retreat, covering themself with their dagger and sword in a cross, we'll repel them and strike them with a decisive blow.

p. 105 §3 The best approach is to catch the enemy from behind, rendering their entire fortification useless by attacking the rear of the retreating men; or by advancing under cover with a spade and shovel until the experienced captain counters this by enclosing the beaten position with another semicircle.

p. 105 §4 If the enemy tries to strike us from behind, we'll detect this and enclose them within our forces with a semicircle, striking them during this manoeuvre.

p. 105 §5 This method of offence and defence continues until the situation reaches a critical point. Here, the defender cuts off and isolates the attacked section, forcing the enemy to devise new attacks.

p. 105 §6 When we find ourselves in such a situation, having been deceived by the
p. 106 §1 enemy who's reached our distance, we'll lose a step of ground and form a new circle, giving the enemy a chance to strike. We'll counter this danger with our usual parries and strikes.

p. 106 §2 To conclude this treatise, we must discuss the assault, which can be executed with valour if, upon arriving with the camp, the general swiftly positions the artillery to dismantle defences and cannons to create a breach, taking advantage of a previously identified weak spot. This surprise attack should be executed promptly to dishearten the defender and prevent them from reinforcing their defences while simultaneously making advances.

p. 106 §3 In this manner, after recognizing the stance and line of the enemy's sword and identifying the weakest part of their body (i.e., the part nearest to us), we'll promptly and decisively attack and strike them.

p. 106 §4 When launching the assault, the Major General arranges their entire army into battle formation in their respective positions, places cavalry at the approaches where the enemy might threaten their forces, and assigns those first tasked with the assault to enter the ditch through its breach, guided by their captains, followed by the ensigns with their standards.

p. 106 §5 They then order the others, who are to support and refresh the assault, to remain prepared, with the Master of the Field overseeing their readiness.

p. 107 §1 Regarding the artillery, they command it to fire at the farthest parts of the breached wall to target the defenders who are positioned behind the breach, aiming to strike the flanks of the attackers. The pieces positioned to dismantle defences should focus on the sides.

p. 107 §2 Simultaneously, the trench is used for musketry and arquebus fire, as the entire trench is armed.

p. 107 §3 With these arrangements, the artillery fire is coordinated and the assault is renewed under that cover. The assaulting troops are encouraged by drumbeats and then advanced. The assault is continuously refreshed by the previously designated forces, who, as commanded, are led by the Sergeant Major to the edge of the ditch to bravely storm the fortress.

p. 107 §4 The defenders organise their forces into battle, with part of them repelling the attackers and using artillery in turn, along with other methods which I'll omit for brevity. The most effective defence is to maintain continuous fire from the arquebusiers, who alternate firing in a disciplined manner.

p. 107 §5 Additionally, soldiers are sent into the ditch to attack the assailants from behind, while the soldiers in the trenches oppose and force them to retreat.

p. 108 §1 In summary, as the Captain General of our army and fortress, when intending to launch an assault, we'll proceed with diligence and resolution. We'll arrange our army into battle formation, ensuring that the exterior view of the enemy's approach is properly managed, focusing on the body rather than just the hand (as others might suggest).

p. 108 §2 This is because the tip of the sword, influenced by even a small movement of the hand, creates a much greater effect. By focusing on the hand alone, we risk missing the optimal angle formed by the imagined line and the line of the sword.

p. 108 §3 We'll direct the powers of appetite (that is, desiring and irascible) to resist obstacles that hinder our goal and instruct the motive power to remain vigilant, ensuring our limbs are ready to act according to the will.

p. 108 §4 Some of these limbs will be involved in the assault, while others will defend those carrying out the offence, similar to how a dagger defends the right side of our body while moving towards the attack. By doing so, we'll achieve our goal, which is victory.

p. 153 §1

TOPPLING AT THE BARRIER[1]

p. 153 §2 In all knightly exercises that are done for sport, the Knight must show agility, skill, and strength of the body. It should be known that this type of spectacle is usually done for two reasons: one is to demonstrate to the spectators the likeness of a military feat with this display; the other is so that each of the combatants can show personal valour. Therefore, among all the games of arms, the Barrier is the most noble, which is nothing but a representation of two battles: first, they come to the push of pike; then they draw their swords and seek to obtain victory with it, which is its aim.

p. 153 §3 Before the melee is formed, combat is done from body to body, and this is done so that each individual not only shows their strength but also allows the spectator to understand the general battle.

p. 153 §4 In all knightly exercises the knight will imitate the truth, as if they were in
p. 154 §1 the actual combat with blunted blades, yet avoiding all extremes, so neither exaggerating nor doing things entirely for show, and striving to make all their movements natural, both of the whole body and its parts, always without any affectation.

1 *abbattimento* – a knocking down, a felling of a tree, a demolition of a building, a reduction, a slaughter, figuratively it can mean depression and melancholy.

p. 154 §2 *In this combat, two types of weapons are used, the pike and the tuck,*[1] *or the sword, of which we'll now discuss in detail.*

p. 154 §3 The pike has three positions, namely: the plumb (which is perpendicular); the middle (which departs from the perpendicular and goes into an angle); and the horizontal (which is found in a straight line from the observing eye to any apparent point in the plane of the horizon).

p. 154 §4 Six things are done with the pike:

p. 155 §1
1. Firstly, one walks the field with it.
2. Secondly, one presents oneself to the battle with it.
3. Thirdly, one walks to deliver the thrust with it.
4. Fourthly, one strikes with it.
5. Fifthly, it's thrown to the ground in order to take hold of the tuck.
6. Sixth and lastly, when departing from the field, one walks with it raised aloft.

p. 155 §2 The parade that a person with the pike performs when appearing in the field should be such that the combatant shows agility in the legs, extending them, and strength in the waist, disregarding the weight of the weapons. I should advise against taking steps that are neither too big nor too small, but rather natural and without any affectation. The pike should be held on the right shoulder with its point below the horizon, and in a way that, drawing a line from the knee, an acute angle is formed with it. Keep the arm straight, not in line with the shoulder, but slightly inclined to avoid affectation.

p. 155 §3 For every time the knight removes the pike from the shoulder to raise it, he should do so in the proper manner, making no movement that's not done at the right time and in coordination. This will be most effective if, when raising the pike, the right foot is lifted, and when lowering it, the right foot is pulled back, with the gaze towards the tip. Likewise, when raising the pike to place it back on the shoulder, the foot is lifted, always following the same rhythm.

1 *stocco* – From the French 'estoc', in English a 'tuck'. Pistofilo (1627) uses the word for short, broad swords used in armoured foot tournaments.

p. 155 §4 The approach to the battle will be done with the body arranged and upright.
p. 156 §1 Since here it's customary to wait for the opponent to be ready, during this time some movements will be made: now with the head, now with one foot, now with the left arm, and now with the entire body. This is done to appear lively and not like a statue. Additionally, the pike will be held perpendicular with the heel near the right foot, and the right arm extended upwards, as shown in this illustration, which depicts the body's proportions as it transitions from stillness to the movement towards our left hand, and the other one as it turns in the passage to the right hand, showing the body font-facing.

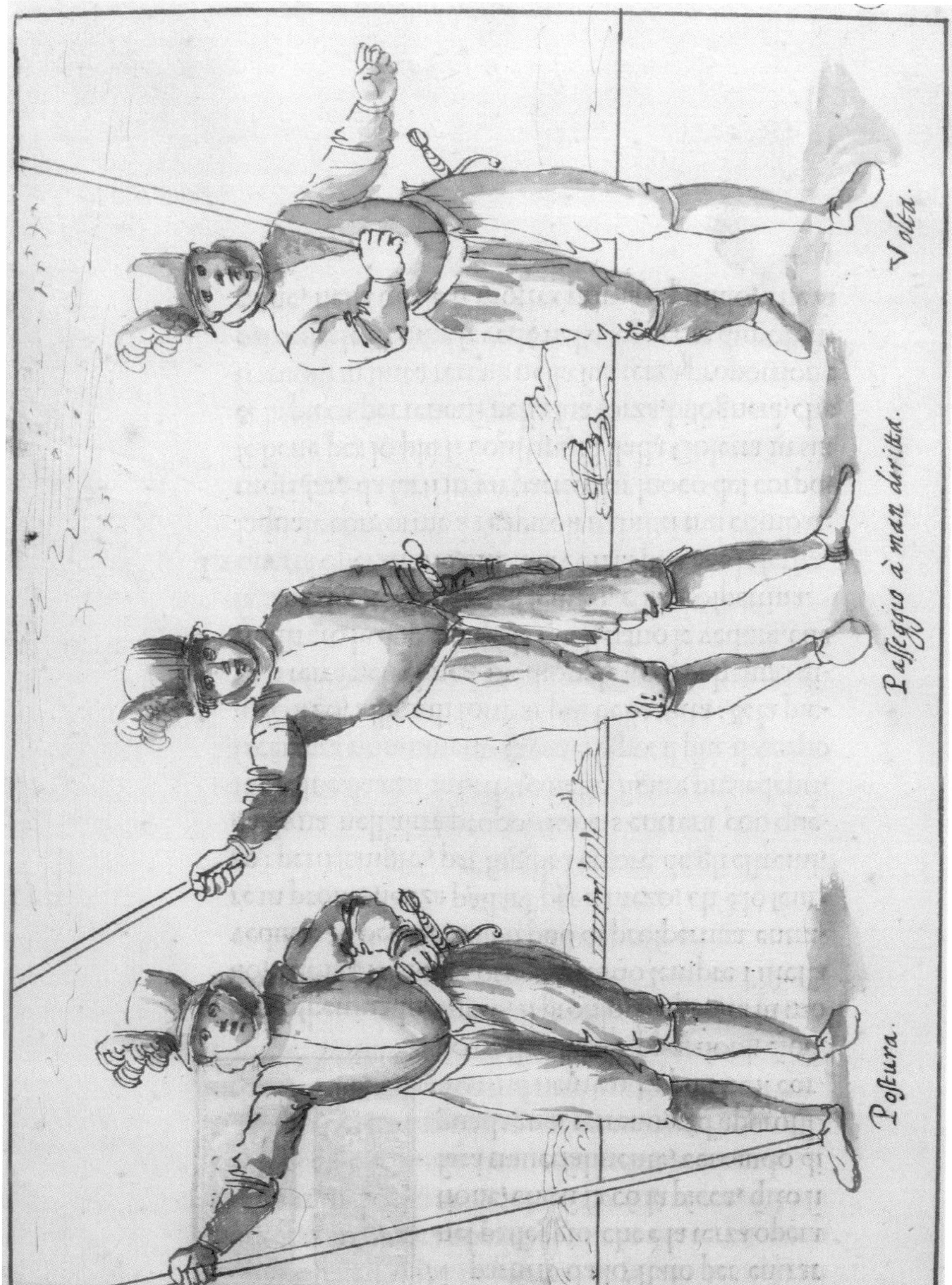

Posture. Passing to the right hand. Turn.

p. 158 §1 When transitioning from the stationary position to entering the movement, which is the third action performed with the pike, it will be done diagonally, seeking to gain ground and approach the enemy. And since our body has three facings (the face-on view, the foreshortening, and the profile), we'll act in a way that keeps the pike aligned with the body in the same view. Since it's not possible to transition directly from the face-on view to the profile without passing through the foreshortening, we'll always, to avoid the error of extremes, move from one facing to another in the order shown in the preceding illustrations. However, we'll strive to keep the body primarily in the foreshortening position to create a more pleasing appearance. The pike will be held in the middle line as demonstrated by the two figures, which represent the views produced by our body: the foreshortening and the frontal view.

p. 158 §2 The fourth action performed with the pike is the strike, which, according to
p. 159 §1 the established rules between the combatants, should be executed at a specific target on the body. Although it mostly targets the throat and above, the pike must be held in a straight line to maintain its effectiveness and in its horizontal third proportion. This can be observed in the following illustration, which first shows the profile formed by the body and the pike when moving to the right hand side, then the second position when the body turns to enter the foreshortening, whether to lower the pike for the strike or to continue the movement, and the third figure represents the strike itself, which is the subject of discussion here.

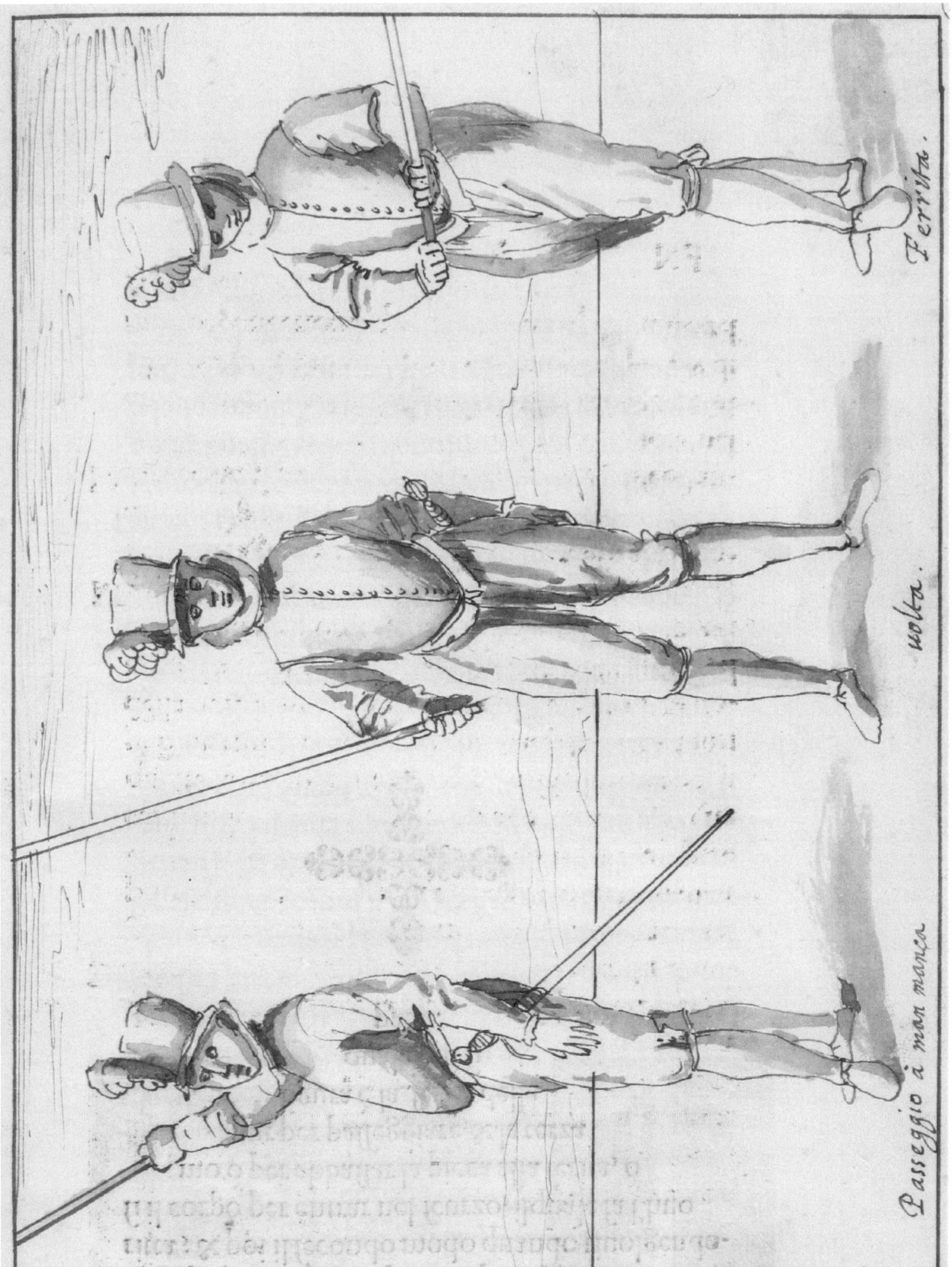

Passing to the left hand

turn.

Armed.

p. 161 §1 Since the true action of the pike is performed with the centre of the body's weight, the knight should exert a force, no more than that, to execute the said strike. This force should be equivalent to three forward steps, which corresponds to the length of a person. Upon passing this distance, the man will initiate another forceful action.

p. 161 §2 Regarding the method of breaking the pike, the best option will be chosen. It can be broken in four ways.

1. In the first, one can place the tip of the pike on the left side to increase its length. It can also be broken by thrusting the right foot to the left side. However, besides the fact that this makes the thrust less visible, it becomes difficult to control the point, as the pike is outside our field of vision. One shouldn't hold the pike in such a way that it's facing downwards. By doing so, the body is not properly positioned to use its strength effectively.

p. 161 §3 But when one becomes the defender, they can better and more reasonably vary their attacks by employing the following methods:

p. 161 §4

2. In the second method, one can strike by moving the arms, extending them away from the body, and attacking by breaking the pike in the opponent's body. It's difficult to parry elsewhere, and all strikes are angular and quite disjointed. However, this is not the true operation of the pike.

p. 162 §1

3. In the third (and this is the very best) one can, with three steps of strength, pass with the right foot, and strike with one hand only.

p. 162 §2

4. In the fourth and last one (which is the one we'll use, once we've walked with a front-facing body), the pike will be lowered little by little, which will be received by the left hand, and then, having reached a distance, with it horizontal, we'll make sure that its end is directly at the point where we want to strike, and withdrawing the left foot close to the right and adjusting the strike, we'll make the movement of three steps: moving the left foot first, and then the right, and then finally we'll drop the weight, and with the movement of the man's strength, end on the left. We'll also keep the right arm at an angle; and the fist close to the belt, to make the strike in a straight line.

p. 162 §3 The fifth operation which is done with the pike, is the retreat after the wound has been made — and this to start the battle again with another pike, or to draw the tuck. Said pike, once broken, will be thrown to the ground by the knight. This throwing to the ground is done in three ways:

1. In the first, backward, or to the side, or over the right shoulder: but this, in addition to the fact that it can't really be done with the pike, is mostly given to the head, and to the seconds,[1] and to those around.[2]

p. 163 §1

2. In the second method, one can throw the pike in contempt towards the enemy. But since it's within the power of the enemy to retaliate in the same manner, in my opinion, it's not advisable to belittle the enemy. Since one will have to deal with the tuck immediately, which holds greater value, the enemy may remember the contempt and respond in kind. For where actions are expected, insults are given little regard.

p. 163 §2

3. In the third method, which we'll use effectively as soon as we've made the wound, we'll raise the hilt of the pike while withdrawing with the left foot. We'll then take a new a new step back with the right foot, letting [the pike] slip through the left hand, and we'll throw it to the ground beneath our right side. This not only looks very impressive but also brings us closer to reality, since once the act with the pike is done, we move on to the melee, and if we let it fall to the ground, we immediately draw our sword.

p. 163 §3 The sixth operation is that the Knight, upon leaving the field after the battle is finished and finding themself in a stance with the pike raised upright, will start moving with the left foot, and then with the right foot next to it, in the same beat, they'll place the hilt of the pike. After taking about five or six steps, properly extending and lowering the pike, they'll rest it on their shoulder and continue on their way.

1 *padrini*, 'godfather' or 'sponsor', was the term used for seconds who assist at a duel or tournament.

2 I'm not sure of Ghislieri's meaning here. Perhaps that there is a danger of hitting bystanders?

p. 164 §1

TREATISE ON THE TUCK.[1]

p. 164 §2 The knight will have three aspects when going with the tuck; agility, strength, and swiftness.

- Agility consists of effectively freeing the strikes, which is greatly enjoyable as it avoids entangling the tuck with that of the opponent.
- Strength, although a gift of nature, requires careful consideration in selecting the best motion for delivering the strike since it allows for greater force.
- Swiftness,[2] being a precious natural gift in matters of arms, ensures that the knight is well resolved in the chosen method of attack, and also that it is achieved in a very short time.

p. 164 §3 Since the nature of motion and cuts (and which of them are more effective
p. 165 §1 according to their descent) has been fully discussed, there's no need to further extend ourselves here. We'll only say that natural motion, aided by the force exerted, will have a greater impact on a resisting opponent. For the true strike will occur when the tuck, along with the arm, forms a straight line, striking with the shoulder.

p. 165 §2 There are several ways one can strike:[3]

- In one, with forehands around; in another, with forehands across, and with redoubled forehands; or one can use an open-handed blow, carrying the weight with the shoulder, making a motion from the centre of the body upwards.
- p. 165 §3 With forehands around, the tuck is not well released and the blow is not the strongest that can be made.
- p. 165 §4 When we strike with forehands across, aided by greater motion and weight of the body, it will be the best method that can be done. For this way, we'll strike in the middle of the temple and the jaw, which is the easiest place to stun the opponent and encourage them to yield.

---

1 *stocco* – From the French 'estoc', in English a 'tuck'. Pistofilo (1627) uses the word for short, broad swords used in armoured foot tournaments.

2 *Presteza* means both swiftness and readiness, and Ghislieri is playing on both meanings here.

3 These are different angles of cut, see p. 32 §2. Here armoured contestants are using blunted swords to batter each other.

p. 165 §5
- Striking with the fifth centre, redoubling, is done for the same reason — so that the enemy is quickly stunned by the short duration of the blows. However, these blows are of little weight, and there's a risk of falling into some disorder.

p. 165 §6
- When the arm is extended with the motion of the body and strikes at the chin, this action is feminine and takes a long time, and it could cause the sword to fall and it also looks ugly.

p. 165 §7 Therefore, it remains that our knight is the first to strike the enemy, since it's
p. 166 §1 always advantageous to be the first attacker. This is done with a forehand across, pivoting around the left foot, so that the circumference is greater. Moreover, the opponent will be deceived by the distance and will be struck while he's in motion.

p. 166 §2 When the knight realises that one of the combatants is disconcerted in delivering their strikes, they could position themself to strike from behind, or be the first to deliver a forehand cut, doing so with less force. Then, when the opponent strikes with a strong backhand, they can assist its descent so that their weight will be leaning over.

p. 166 §3 It should be noted that when striking the opponent, if we see them disconcerted by the hit, we'll execute the same attack. Similarly, if we observe ourselves being disconcerted by a forehand, we'll redouble the forehand. The same applies to backhands.

p. 166 §4 In the event that the swords become entangled while moving our hands, we must be mindful not to pull towards ourselves, but towards the ground. During this time, it will be necessary to place the left knee below the barrier and assist with both hands.

p. 166 §5 If, as often happens, the opponent yields and their body comes over the barrier, we'll make them fall towards us with the left hand — that is, by pulling them over by their helmet.

p. 166 §6 There are two ways to make the opponent's sword fall:

p. 167 §1
1. In the first, when they strike with a cut, we'll strike their arm with a cut as well.
2. In the second, when they intend to perform a backhand, we'll place the point of our sword on their right shoulder, forming a line with it.

p. 167 §2 But to come closer to practice, the knight must know the rules and the regulations established by the Masters so that they can act accordingly. So, following the order of the pages, seconds, and drummers, they'll enter the field. I won't say anything more regarding this order since it's well-known.

p. 167 §3 While confidently walking through the field, when they reach the presence of the Ladies, they'll raise their pike and pay homage to them in the manner previously described. Then, continuing on, they'll do the same in front of the Judges, albeit with less humility.

p. 167 §4 Once they've arrived at the designated location to engage in the game, the
p. 168 §1 knight, raising the pike in the appropriate manner, will present themself ready for the battle and demonstrate agility in their arms. When lifting the right foot, they'll simultaneously raise the pike high with the right hand, while the left hand will grasp it at a respectable distance from the right. Then, pulling the right foot back, they'll pay respect to the Ladies. Finally, as the bout concludes, the same foot will complete its lateral movement, bringing the pike to the descent across.

p. 168 §2 Then, as the knight brings back the right foot, they'll bring back the pike, passing with the foot and the pike face-on. They'll grip the pike with the right hand alone, with the body in profile and the pike in the specified position. As the left foot rises, the pike will be raised high, and in this act of reverence with the left hand, the pike will be lowered. Continuing in the walking stance with the left foot and the body in profile, the pike will be raised high, always maintaining it in the middle position. They'll execute the same movements with the foot and the hand.

p. 168 §3 Then, when the knight reaches the end of the walk, they'll deliver a strike, observing what the opponent does. If they see that the opponent hasn't yet engaged in combat, they'll return to the same walking stance, making a slight movement of the body to indicate that their willingness to fight is ready. But when they finally see the opponent coming into combat, following the previously mentioned method, they'll go to strike them. After delivering the blow, they'll withdraw, retrieving the other pike. They'll continue this way, and once the last pike is broken, they'll immediately switch to the tuck from below.

p. 168 §4 For to ensure that everything proceeds in harmony, it must be known that
p. 169 §1 instead of the retreat, which is done by raising the left foot and then retracting it, the pike is raised. Then, while retracting the right foot, the same pike is thrust below the right flank. In this action, when raising the left hand, the pike is raised, and in the step forward, the same pike is thrust below the left flank, below the sword, assisted by the left hand. Then, while retracting the right foot, the hand is simultaneously brought to the tuck. And by continuing the retreat, the following actions are performed in the manner that follows.

p. 169 §2 A reverence will be made with the right foot and the body turned towards the Ladies, holding the tuck diagonally above with the left hand. Then, stepping forward with the right foot in that motion, the tuck will be raised high, turning towards the Judges, and with the left foot making a gesture of reverence. Then, swiftly passing again with the left foot, the tuck will be regained by its far end with the left hand from above. While regarding the barrier, the knight will present themself two steps away in a position to strike with the left foot forward, and as he approaches, holding their breath with all their might, they'll strike the opponent with a diagonal cut. By following up with the backhand, they'll conclude their attacks. Once completed, they'll withdraw the right foot one step back, and with the tuck raised high in a striking position, they'll make a bold movement of the head.

p. 169 §3 One might perhaps doubt whether a reverence should be made with the
p. 170 §1 tuck because, once the pike is broken, the battle is immediately engaged with the tuck. Therefore, I respond that, since the pike and the tuck are different weapons, and since each of them is rewarded, and because making a reverence gives time to the spectators to prepare for the rest, it should be done, although it may matter little to the battle.

p. 170 §2 The knight will then retreat with the body facing forward, using the steps mentioned above, all the while brandishing the tuck. Once they reach their second, they'll immediately sheathe it to give way to the other combatants.

p. 170 §3

TREATISE ON THE MELEE

p. 170 §4 There are broken, as agreed upon, one or three pikes in the manner mentioned above. Once this is done, p. 171 §1 they engage in tuck combat, and in order to counter-position, one must first place oneself on the right side, so that they can push their companions with their left side. Also, they should be reminded to always strike from the right, since striking from the left wouldn't work well due to the obstruction of their companions, and all blows (both from enemies and friends) would fall upon that arm. Thus, this is all that I need to say about the combat that takes place in the barrier.

p. 171 §2 However, I must still advise the Seconds that when one hands the pike to one's combatant, one should first take their right hand with one's left hand. Then offer them the same pike, instructing them to hold it in their right hand with ease, so that the combatant has no reason to use both hands to adjust it properly in their right hand, which is an unsightly thing to see.

p. 171 §3 The drummers should beat the call when they arrive at the field. When they enter formation and the parade begins they should beat the call again. When the pike is lowered to strike (the battle), and when the pike is broken (the retreat), so it should be done from time to time as needed. The same order shall be followed when fighting with the tuck.

p. 171 §4 I haven't wanted to treat here certain vain flourishes which can sometimes be done for showmanship, p. 172 §1 because the invention of these belongs to the Masters of the Art: for when one follows the reasons and rules that I've given above, they can create and compose them according to their own style. Nevertheless, they should strive to remain as much as possible on the side of plausibility.

p. 172 §2

TREATISE OF A FOOT COMBATANT AGAINST A HORSEMAN

p. 172 §3 When a man on foot intends to oppose someone who fights on horseback, he must act in accordance with the nature of the horse, which, being built with a long body supported on four legs, walks straight ahead as its natural movement. Any other kind of movement will be unnatural for it. Therefore, if the horse wants to move its front or right side forward or to the left, it will pivot on the hind legs as the centre. And conversely, when it moves the hind legs, it pivots on the front legs as the centre.

p. 172 §4 Since it's inherent in all creatures to instinctively flee from things that pose
p. 173 §1 harm to their well-being (for they're driven by their innate desire to preserve themselves, as bestowed upon them by nature), it follows that when a horse is struck in a certain part of its body, it instinctively veers away from the source of the injury. Even if it's spurred on its flanks, when it sustains a more substantial blow in a different area, one that surpasses the discomfort caused by the spurs, it will instinctively move away from the side where it experiences the greater attack.

p. 173 §2 Thus, when facing an opponent mounted on a horse, the enemy will attempt to achieve the opposite by forcing the horse to flee or by disarming the rider through targeted strikes towards their right hand with a sword. However, the foot combatant will strategically evade such attacks by constantly positioning themselves to their opponent's left, and aligning their own right side towards the adversary. By doing so, they not only distance themselves from the enemy's offensive manoeuvres but also move closer to engage the opponent effectively, as depicted in the accompanying diagram:

p. 175 §1 Likewise, because the horse can only turn towards us naturally by going straight, we'll take care to always present ourselves on the horse's direct line. This way, we'll obstruct the enemy's view, who'll find themself on horseback. And standing in a moderate step, united in strength, with the sword extended beyond our body and with the cape gripped in our fist in such a way that we can quickly free ourselves from it if needed, as soon as the horse reaches a suitable distance, we'll deliver a cut to its muzzle or a thrust at the same time, stepping away with the right foot followed by the left, towards our right side. So we'll strike with a backhand across the reins. Then, positioning ourselves in the centre of the circumference that the horse forms when it turns its body in a circle, we'll remain here without abandoning the enemy, continually striking either the horse or the rider.

p. 175 §2 Furthermore, at the moment when the horse reaches the distance, one could also frighten it with the cape or throw it over its head while making the same steps and attacks.

p. 175 §3 It's also advisable to use the left hand on the bridle; or with the same hand give him a lift beneath his foot, and thus unhorse him.

p. 175 §4 Should the horse counter-position, one should immediately deliver a cut to its hocks.

p. 176 §1

TREATISE OF ONE ON HORSEBACK

p. 176 §2 *who fights with the sword against another, also on horseback.*

p. 176 §3 The same principle that we use to gain the straight line while on foot, we'll also use while on horseback. This is even easier because the horse's movement is more steady than our own when we were on foot, and the horse serves the rider primarily for the convenience of their legs.

p. 176 §4 Since the horse has a long body, it forms a line: when it moves diagonally, one end remains still while the other forms a circle. Thus, when the horse moves forcefully, pivoting on its hind legs, it creates a circle with the rest of its body, with the centre of the hind legs as the pivot point.

p. 176 §5 Yet, just as on foot, by gaining the diameter of the circle, we gain the straight
p. 177 §1 line and we force the enemy to pass through it. So when on horseback, we'll always gain the diameter of the circle. Furthermore to provide some understanding of this, let's explain.

p. 177 §2 One must always approach the enemy head-on (that is, face to face). This will facilitate our goal even more, even if the opponent tries to evade. This way, we can easily maintain the centre with our horse, forming a much smaller circumference than the enemy. We must also ensure to stay in a straight line. However, assuming the enemy comes close, once they reach a certain distance from any side, when the horse turns its head, we'll move straight in and keep the sword extended with the hand in front of the saddle, crossed in a way that the point faces the left ear of the horse. In this position, we're ready to parry and strike, avoiding any other blows while on horseback except for these two: a thrust from below towards the length of the horse's neck and a backhand thrust along the same path, as shown in this diagram, where the drawn line represents the posture and the sword represents the strike.

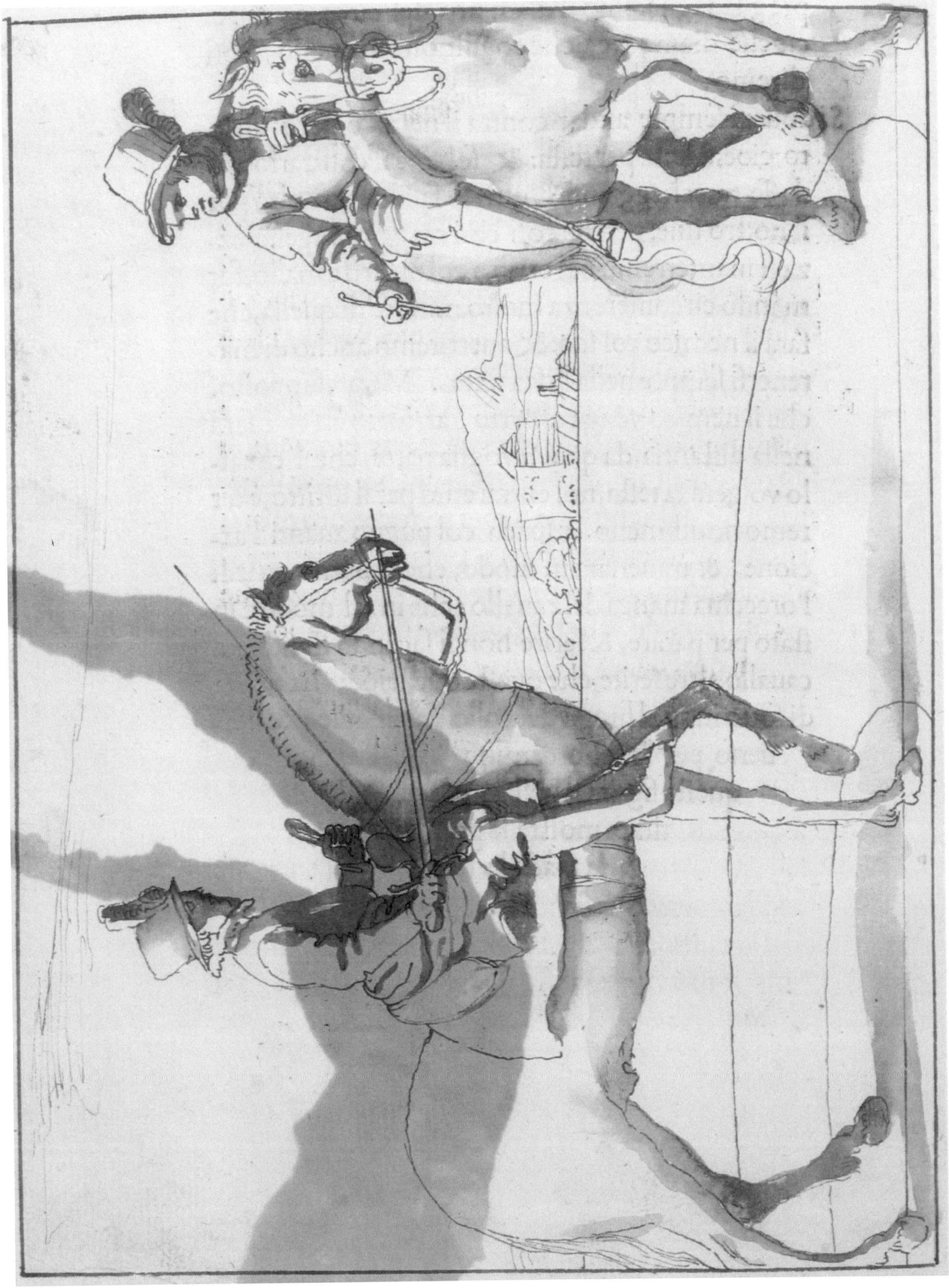

p. 179 §1 In case the opponent turns around and charges, we'll follow them in a straight line with the head of our horse towards the tail of theirs. For when they decide to turn, we'll move into the diameter. Furthermore, when we've a good horse underneath us, we can control it by placing it under the right hand. This way, we can strike at the head or the flank of the horse.

p. 179 §2 Then again, if the knight finds themself on a horse that isn't suitable for the encounter, they may decide to avoid the confrontation while still intercepting the opponent who approaches in a diagonal manner. By putting their horse into flight, they'll extend their right arm in a straight line and with force. By doing so, they'll always find the opponent, who'll have their arm out of force. From this position, it will be easy for the knight to make the opponent drop their sword or thrust.

p. 180 §1 *The exercise of jousting with lances was invented both for leisure and to inspire the Knights, as well as to prepare them for actual combat. Through these practices, they aim to become skilled in these operations, so that they may be experienced when it comes to the real thing. To facilitate this method, the running at the Ring was invented. Although it's also customary to break lances in the Quintain or to practise without it. However, since all these exercises serve a single purpose, which is to know how to wield the lance effectively in open field and in war, and because among all these games, running at the Ring is the most common and widely practised, we shall discuss it briefly.*[1]

p. 180 §2 Firstly, there are six circumstances that contribute to the successful jousting with lances.

1. The first is holding the lance firmly on the thigh.
2. The second is lifting it with the appropriate posture.
3. The third is smoothly stopping it.
4. The fourth is lowering it at the right moment.
5. The fifth is striking at the opportune moment.
6. And the sixth is recovering it as necessary.

p. 180 §3 In the first circumstance [holding the lance firmly on the thigh], the lance can be held on the thigh in three ways:
p. 181 §1

1. Firstly, the Knight should be aware that when he's armed, he must hold it between the thigh and the stirrup, as the equipment does not allow it to rest directly on it. In the first way, the lance is held with the point towards the Knightly's right side, in the Albanian style, and this is called an open lance.
2. In the second way, it's held with the point towards the left hand
3. To avoid extremes, we should hold it in the middle so that it aligns with the body. We hold it in the middle with the arm straight, forming an angle with the elbow. It should neither be too high nor too low, but in a way that allows us to comfortably exert force, and this will be the third manner.

---

1 This section apparently draws from a dialogue in Giovanni dall'Agocchie's *Dell'Arte di Scrima Libri Tre* (1572):
https://pieceofheaven1951.blogspot.com/2020/02/sometimes-massive-research-projects.html

p. 181 §2 The second circumstance, which is lifting the lance from the thigh, can be done in six ways.

p. 182 §1

1. In the first, it's lifted with the fist facing upwards, gradually stopping and supporting it by the wrist.
2. In the second, it's lifted by extending the arm downwards along the right side.
3. In the third, it's lifted by widening the fist away from the body with the arm extended in the Albanian style.
4. In the fourth, it's lifted upwards and immediately stopped at the Stradiot position.[1]
5. In the fifth, it's lifted with the arm extended along the right side, slightly bent, and the hand turned inward.
6. In the sixth (and this is the method we'll now discuss, since it's the best), it's lifted from the thigh while keeping the fist on the same plane, and by supporting it with the wrist, the point of the lance is positioned in line with the Ring, and the arm is slightly bent, which makes it easier to sustain and allows for easier stopping. This method follows the practice in warfare.

p. 182 §2 The third circumstance, which is stopping the lance, can be done in four ways.

1. In the first, it's stopped at the beginning of the charge.
2. In the second, when the horse is in motion.
3. In the third, when the knight approaches their adversary at a distance of a lance's length.

These three methods not only look ungraceful, and although the last one can still be done, when running with the lance couched, the lance easily wavers due to the motion of the horse, and it should be held steady, otherwise it produces a very unattractive effect. The manner of stopping can be done with the hand turned upwards when running without a weapon, and turned downwards. This method is considered the best since it's a natural motion, unlike the others which are forced.

4. The fourth way is to stop at the same time the lance reaches the straight line in the thrust.

---

1 Stradiots were mercenary Balkan horsemen. The light cavalry were recruited from the C15th until the middle of the C18th, mainly by Southern and Central European states.

p. 182 §3 The fourth circumstance, which is lowering the lance, is done in four ways.

p. 183 §1
1. In the first, as soon as it's stopped, it's lowered.
2. In the second, when the Knight approaches the enemy, it's slightly lowered.
3. In the third, it's lowered and strikes at the same time.

The first shouldn't be done in any way, the second is more acceptable, and the third makes a beautiful gesture when it can be done, but it's difficult.

4. The fourth, which is the best of all, is done by gradually lowering the lance, keeping it collected in such a way that it finishes the downward movement in the thrust.

p. 183 §2 The fifth circumstance, which is the strike. If we want to be sure of the point [the target], then as soon as we set off we'll position the lance crosswise so that its tip is aligned with the point that we intend to strike. Our line of sight, passing outside the lance, will reveal the point or the target.

p. 183 §3 The sixth circumstance, which is the recovery of the lance, can be done in two ways.

1. In the first, it's lifted up and placed back on the thigh.
2. In the second, the heel of the lance is thrown behind the right thigh, causing its tip to face backward. Then, when the horse is halted, it's placed back on the thigh.

Furthermore, I'd distinguish these two methods: when running at the ring, after the strike is made, the fist should be raised up, and when the horse is prepared, it should be placed back on the thigh. This is because the intact lance should be held in the position to strike. But if the lances were to break, I'd recommend that the broken piece be placed behind the right thigh.

p. 183 §4 Now, drawing a straight line that intersects the Ring with the line of the
p. 184 §1 course, two parallel lines will be formed. The line of the lance, when striking the Ring, will form an acute angle with the given line, as shown in the demonstration below. Therefore, the lance should always be held in such a proportion that it's ready to strike with the least possible motion. The Ring will be placed 3' away from the line of the course and to the left hand side of the knight who'll run. Furthermore, it will be positioned 6' above the ground.[1] However, in general, each person who wishes to run should first determine the height of the Ring, which should allow the knight to touch it with the tips of their fingers. In this way, the runner will become better accustomed to running with the lance held high, so that when they subsequently encounter their opponent, either in the lists or in open field, they'll easily strike at the head, which is the most notable target.

1 This local unit of measurement might be about 25% larger than the modern unit. See: https://pieceofheaven1951.blogspot.com/2020/02/sometimes-massive-research-projects.html

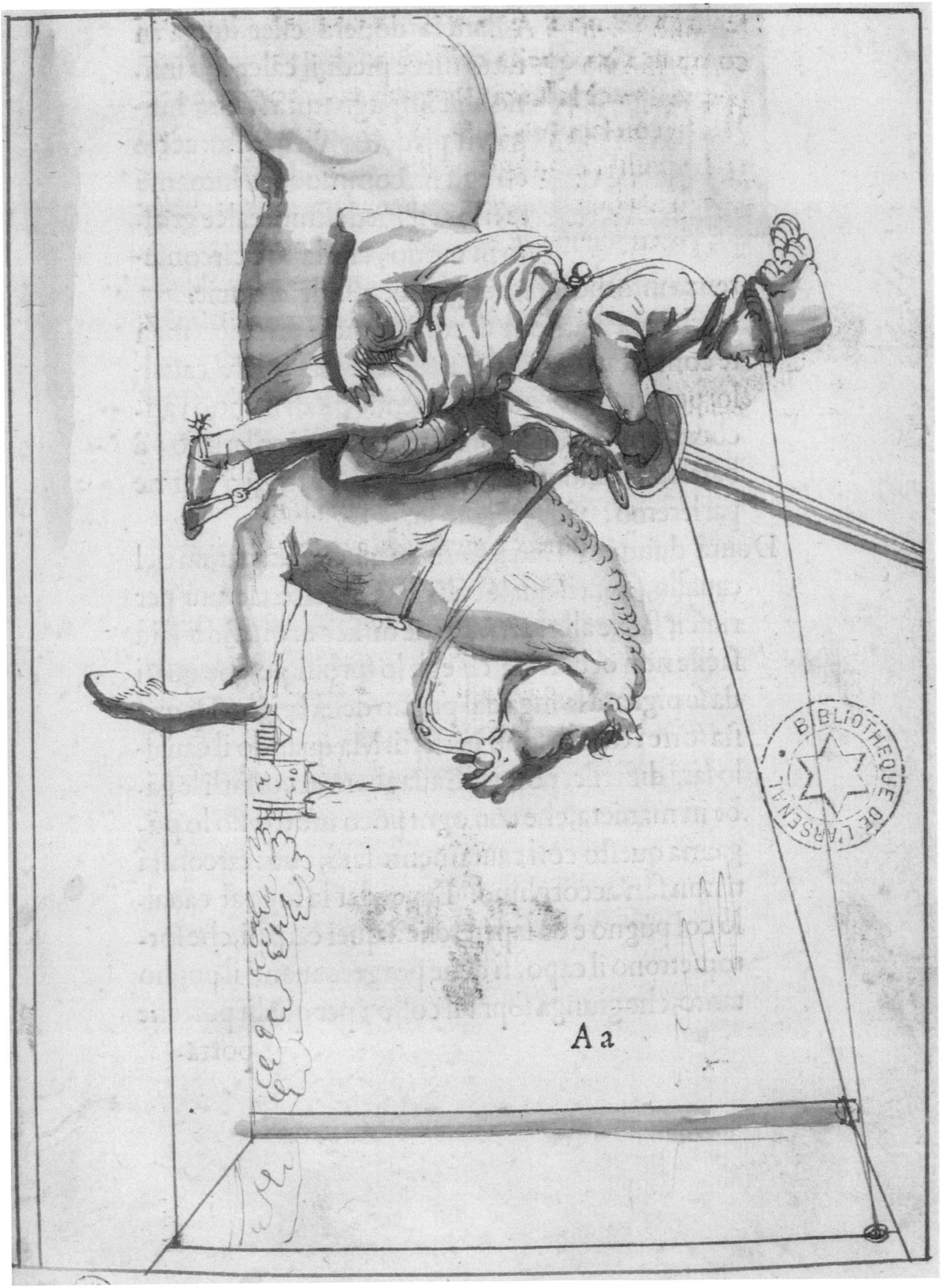
Aa
BIBLIOTHEQUE DE L'ARSENAL

p. 186 §1 The lance should be 10’ long in total, and its shaft up to the grip will be 1¼’ long, in order to provide comfort. Lastly, the same shaft will be made thick in such a way that its circumference around the grip is six inches.

p. 186 §2 Since this action of jousting primarily relies on the rider’s ability to handle the horse, as almost everything depends on it, although this is the profession of a horseman, we’ll also discuss it enough for our purpose here.

p. 186 §3 The rider must therefore understand the nature of the horse upon which they'll
p. 187 §1 joust — whether it will be easy or difficult to engage in the charge. There’s be no need to spur it when it'll be easy, since it’ll almost take flight from the pressure of the reins. Horses of this kind are perfect.

But when the horse is difficult, the rider may approach its legs in such a way that with minimal movement, he pricks it, but he must do so cautiously so that others do not notice. And to urge the horse forward with the reins, it should be noted that for horses that lower their head, the hand should be extended forward so that it reaches above the neck. For those that hold their head high, the hand holding the reins should be lowered. However, it’s important to note that the rider’s will shouldn’t be so strong that the horse has to lean forward on its left shoulder, since this looks very unattractive and should be avoided at all costs. This is especially harmful when jousting, since it exposes the left side to the enemy, shortens the lance, and leads to many errors which can be avoided. To appear more elegant and graceful on horseback, the rider will keep their body upright and won’t sit too comfortably in the saddle. They'll turn their gaze, passing between the horse’s ears, and by squeezing their knees, they'll allow their legs to hang naturally and perpendicularly. In this position, they'll show such strength with their flat foot that the tips [of the feet] align with the horse’s shoulder, leaning neither downwards nor upwards.

p. 187 §2 Thus, the Knightly rider will divide their charge into five parts, and these divisions must be kept distinct in their imagination, while in practice they’ll be connected in such a way that onlookers can’t distinguish them as separate actions. These parts are:

1. Raising the lance
2. Gathering it
3. Lowering it
4. Striking
5. Recovering.

p. 187 §3 Once the Knightly rider is prepared for the charge and positioned correctly,
p. 188 §1 they'll first raise the lance from their thigh. At that moment, they'll spur the horse forward.

1. In the first part, as they lift the lance from their thigh as we've described.
2. In the second part, they'll advance while raising their arm, forming an angle.
3. Upon reaching the level of the shoulder in the third part, they'll arrest [their arm] by lowering the lance.
4. In the fourth part, they'll strike in such a way that the end of the lowering hits with the lance.
5. In the fifth part, they'll recover the lance onto their thigh using the aforementioned method.

p. 188 §2 The knight can also practise running on foot while armed only with a breastplate, as this will improve their wrist technique. Once they become proficient in this exercise, they can confidently mount the horse and engage in running exercises, either aiming for the ring or the glove. After mastering these exercises, they can proceed to break lances.

p. 188 §3 Once the riders have demonstrated their ability to run with proper technique, proving themselves to be skilled and capable, they can proceed in various ways. They can use the lance in an open position, or reverse the position of the handle, placing it in their left hand. When breaking lances, they can do so with a leap in the air, or by striking from behind the neck. They can also position it on their shoulder in the opposite direction, and while running, throw it and catch it in its usual place. Another option is to hold it by the handle and break it from top to bottom.

p. 188 §4 Lances are also often broken on the ground, which is a very common method,
p. 189 §1 or in Quintain. The Quintain (to explain a little more) is set up to resemble a wooden man on horseback, placed on a cart so that it moves, allowing the Knightly rider to practise better when running towards it.

p. 189 §2 For this practice of Quintain a barrier will be set up. It will be 200' long and 5' tall. A counter-barrier will also be constructed, which will be one 150' long and 2½' tall from the ground. It will be curved inward towards the barrier so that it does not collide with it when the legs are spread to strike the horse. Specifically, the distance above will be 3½', and below it will be 4'.

p. 189 §3 Furthermore, to ensure that fewer lances are broken during practice (to reduce expenses), it can be arranged that the shaft of the lance is 4' long, with a well-fitted cannon at the top.[1] It should be designed in such a way that the remaining part of the lance, 6' long, can be inserted into it with a fitting joint.

p. 189 §4 It's also known that when one rides in an open field one must be careful not to strike the horse with the left spur, which would cause it to shy from the collision. The fist should be clenched to prevent the rest from causing damage to the hand if the lance were to rise and pull forcefully upon impact.

p. 190 §1 For the final consideration, I also say that when one rides, not for play, but to
p. 190 §2 engage in combat with sharp blades,[2] one must know that upon approaching the enemy,one should press the horse against them to strike, and this should be done in a straight line as much as possible. One should aim for the left side or attempt to pin the thigh, which is the most vulnerable part.

p. 190 §3 *THE END.*

1 The 'cannon' here being the tubular half of a socket-and-spigot joint.

2 *col ferro ammollato* – literally 'with loose iron' or 'wet iron'.

Appendices

Extant Copies of *Regole di Molti Cavaglierreschi Essercitii*

France

Bibliothèque nationale de France, département Arsenal, RESERVE 4–S–4348

This copy was digitalised in 2018, and was used as the basis for this transcription and translation.

https://gallica.bnf.fr/ark:/12148/bpt6k1518476n

United Kingdom

Kelvingrove Museum, R. L. Scott, Glasgow, UK. E.1939.65.1566

https://collections.glasgowmuseums.com/mwebcgi/mweb?request=record;id=240892;type=101

This is the primary copy used by Dr Anglo. It is dedicated to Ranuccio Farnese and has 29 illustrations.

The bookplate marks it as EX LIBRIS M. A. PRINCIPIS BVRGHESII. Ghislieri first served the church during the sede vacante when Camillo Borghese became Pope Paul V. Francesco Borghese, Paul V's brother, was appointed head of the Church's military forces for the entire period Ghislieri worked for them.

Unfortunately a handwritten note on the verso of the last page of the dedication has been scribbled out.

British Library, London, UK 1570/2850

This copy is dedicated to Antonio Pio Bonelli, reproduced in a subsequent appendix. Apart from two straight lines drawn in the geometry section, it otherwise lacks illustrations or marginalia.

Italy

Museo dell'Arte Marziale, Botticino, Brescia, Italy, M.A.M. Ghisliero MS (1585)

An illustrated manuscript copy of Ghislieri's book is attested to in Gotti's collection.

Otherwise there appear to be at least five surviving Italian print copies:

https://edit16.iccu.sbn.it/resultset-titoli/-/titoli/detail/CNCE20863

Biblioteca Universitaria di Bologna, Bologna, Italy. BUB, RARE C 1

This copy is missing the title page and dedication, but still mentions Ranuccio Farnese. The first page also has cursive writing.

The manuscript illustrations are not drawn directly into the book, but rather cut from another copy of the work and glued onto the blank pages in the correct places. It lacks illustrations after the sword and cape.

Biblioteca Nazionale Vittorio Emanuele III, Naples, Italy. S.Q. 35. A 0055 / A01 1456963

This copy is dedicated to Ranuccio Farnese, and appears to have been owned by Farnese.

https://polosbn.bnnonline.it/SebinaOpac/resource/regole-di-molti-cauagliereschi-essercitii-raccolte-dal-capitano-federico-ghisliero-per-seruitio-del-/NAP1617535

Archivio di Stato, Padua, Italy, ADO.CINQ.255

This unillustrated copy is dedicated to Ranuccio Farnese, and lacks any marginalia. It belonged to the Arrigoni degli Oddi library from the mid-C19th.

https://galileodiscovery.unipd.it/

permalink/39UPD_INST/prmo4k/alma990033098800206046

Biblioteca Reale di Torino, Turin, Italy P.M.144 / COR–3544

This copy is dedicated to Ranuccio Farnese.

http://www.librinlinea.it/search/public/appl/dettaglio.php?bid=TO0E036479#ancora_23

Biblioteca Palatina, Parma, Italy, AA V.25129 / PAL 43150

This copy is dedicated to Ranuccio Farnese.

https://biblioteche.parma.it/SebinaOpac/resource/regole-di-molti-cauaglierreschi-essercitii-raccolte-dal-capitano-federico-ghisliero-per-seruitio-del-/PAR0983252

United States

Folger Shakespeare Library, Washington, DC, United States, U101 .G5 1587 Cage

This copy is dedicated to Ranuccio Farnese. It contains marginalia and a crude illustration. p. 135 contains a handwritten list of names:

Antonio Capello
Lo[re]ñza Priuli
Marc'Ant.° More[...]
Alessed[...] Lazera

Antonio Capelli, Lorenzo Pruili, and Marcantonio Morosini were all Venetian agents during the period.[1]
https://catalog.folger.edu/record/75529

---

1 https://www.treccani.it/enciclopedia/antonio-cappello_res-d2459f01-87e9-11dc-8e9d-0016357eee51_(Dizionario-Biografico)/

'Venice: July 1581', in Calendar of State Papers Relating To English Affairs in the Archives of Venice, Volume 8, 1581-1591, ed. Horatio F Brown (London, 1894), British History Online https://www.british-history.ac.uk/cal-state-papers/venice/vol8/pp13-16 [accessed 2 May 2025].

https://www.treccani.it/enciclopedia/marcantonio-morosini_res-3386080b-07d0-11e2-8c38-00271042e8d9_%28Dizionario-Biografico%29/

Historiography

Given the exciting life of it's author, it's remarkable that so little attention has been paid to Ghislieri's book. Fencing historians in the English speaking world have always been aware of the work, for it was listed in various bibliographies, such as Castle's in *Fencing, Boxing, Wrestling* (1889),[1] Thimm's fencing bibliography (1896)[2] and, in Kelso's C16th gentleman's bibliography (1929).[3]

In terms of modern academic work, the Ghislieri's text was championed by Professor Sydney Anglo, who gave the first English description in 1994, in an article in *Apollo*.[4] Anglo discussed it again in *The Martial Arts of Renaissance Europe*, a seminal text to the modern historical fencing movement.[5] In 2007, Anglo referenced the work in an article in *Dance Research*.[6] An article on the historical fencing website *HROARR* by Conti mentions Ghislieri and picked up on Piccolomini's comments about other fencing instructors.[7] Mondschein briefly mentioned Ghislieri in a 2016 article, noting for the first time in English the connection to Galileo Galilei.[8] Tobias Capwell also mentioned Ghislieri in his contribution to a work on Charles V in 2021.[9]

---

1 Pollock,Walter H; Michell,E B; Armstrong, Walter, *Fencing, Boxing, Wrestling*. London, 1889, p. 281

2 Thimm, Carl Albert, *A complete bibliography of fencing & duelling, as practised by all European nations from the middle ages to the present day*. London, 1896, p. 163.

3 Kelso, Ruth, *The doctrine of the English gentleman in the sixteenth century, with a bibliographical list of treatises on the gentleman and related subjects published in Europe to 1625*. Urbana, University of Illinois Press, 1929, p. 214.

4 Anglo, Sydney, 'Sixteenth-century Italian drawings in Federico Ghisliero's *Regole di Molti cavagliereschi essercitii*', *Apollo*. 1994, Num. 393, Vol. 140, pp. 29–36. Thanks to Tea Kew for helping with access.

5 Anglo, Sydney, *The Martial Arts of Renaissance Europe*, Yale University Press, 2000.

6 Anglo, Sydney, "The Barriers: From Combat to Dance (Almost)", *Dance Research*. 2007, 25:2, pp. 91-106
doi: 10.3366/drs.2007.25.2.91

7 Conti, Andrea, *'Meyerozzo: The influences of the Bolognese method in German Rappier'*, 2015: https://hroarr.com/article/meyerozzo-the-influences-of-the-bolognese-method-in-german-rappier/

8 Mondschein, Ken, "The Italian School of Fencing: Art, Science, and Pedagogy.", *A Companion to Medieval and Early Modern Fightbooks*. ed. Timothy Dawson, Daniel Jacquet, and Karin Verelst. Leiden: Brill, 2016.

9 Capwell, Tobias, "Chapter 11. "Theatre of War: Armoured Combat at the Festivals of Binche", *Charles V, Prince Philip, and the Politics of Succession*. January 2021, 243-320, p. 292.

More thorough examination of Ghislieri's life was undertaken on the continent in the early C21st. 2000 saw a thoroughly researched article on Ghislieri published in the Italian Biographical Dictionary by BRUNELLI.[1] In 2005, BRUNO published an article on Ghislieri's archival records.[2] A Hungarian research paper full of biographical details was written in by SZOVÁK in 2021, following up a 2018 article.[3]

In 2018, the Bibliothèque nationale de France uploaded a digitalisation of their copy of *Regole di molti cavagliereschi essercitii*. In 2019, the relationship between Ghislieri and dall'Aggochie was picked up by an equestrian historian from Texas.[4] The digitalisation was noticed by VINCENT on the *Ensis Sub Caelo* blog in 2020.[5] KATE SOKOL also began working on a translation around the same time.[6]

In August 2023, I published a first draft translation, although I had missed that GOTTI had acquired a manuscript of Ghislieri's work.[7] In March 2024, Doctor NICOLA BOYD published another translation, which was kindly donated to the Wiktenauer project.[8]

1 BRUNELLI, GIAMPIERO, "GHISLIERI, Federico Fabio", *Dizionario Biografico degli Italiani*. 2000, Vol. 54.

2 BRUNO, ANDREA JR. 'Federico Ghislieri soldato, trattatista, inventore nei documenti dell'Archivio di Stato di Torino', *L'architettura degli ingegneri. Fortificazioni in Italia tra '500 e '600*. Ed. Angela Marino. Roma, 2005. pp. 43-52.

3 SZOVÁK, M. 'Federico Ghislieri hadmérnök szakírói tevékenysége különös tekitettel a tizenöt éves háborúra', *Hadtörténelmi Közlemények*, 134. évf. 1. sz. (2021.) pp. 73-102.

See also SZOVÁK, M. 'Espugnazione per negati: La fortezza di Kanizsa al centro di un trattato secentesco di arte militare.' *Verbum – Analecta Neolatina*, Vol. 19 (1-2), 2018, pp. 119-127. Retrieved from: https://ojs.ppke.hu/verbum/article/view/320

4 https://pieceofheaven1951.blogspot.com/2020/02/sometimes-massive-research-projects.html

5 https://blog.subcaelo.net/ensis/federico-ghisliero/

6 https://timetravelerswardrobe.wordpress.com/2020/05/11/ghisliero-translation-part-1-of/

7 https://msmallridge.wordpress.com/wp-content/uploads/2023/07/ghisliero-1587-rules-of-many-knightly-exercises.pdf

GOTTI, ROBERTO, "The Dynamic Sphere: Thesis on the Third State of the Vitruvian Man." *Martial Culture and Historical Martial Arts in Europe and Asia*, 2023: pp. 93-147, Ed. by DANIEL JAQUET; HING CHAO; LORETTA KIM. Springer. doi: 10.1007/978-981-19-2037-0_4

8 https://catalogue.nla.gov.au/catalog/10024006

Timeline

1532	– Pope Clement VII elevates Alessandro de' Medici to Duke of the Republic of Florence.[1]
1536, May 23rd	– Achilles Marozzo publishes a fencing book in Modena, with woodcut illustrations.[2]
1537, Jan. 6th	– Alessandro de' Medici is assassinated, leaving behind no legitimate heirs. Cosimo de' Medici I, a seventeen-year-old distant cousin, inherits the title.[3]
c.1538–44, Aug. 11th	– An undated letter from one of Cosimo's servants to Pier Francesco Riccio discusses the artist Giulio Clovio moving to the Tuscan court from Rome. It mentions in passing a "Francesco di Sandro", probably Altoni.[4]
1543	– Silvio Piccolomini is born.[5]

1 Rudolf von Albertini, *Firenze dalla repubblica al principato: storia e coscienza politica*. Einaudi, 1995, p. 200.

2 Marozzo, Achille, *Opera Nova*. 1536, Moderna, e.g: https://opacplus.bsb-muenchen.de/title/BV001451632

3 Fletcher, Catherine, *The Black Prince of Florence: The Spectacular Life and Treacherous World of Alessandro De' Medici*. Oxford University Press, 2016, p. xvi.

4 Archivio di Stato di Firenze, *Mediceo del Principato*, Vol. 1171, f. 532r, & Brunelli, Giampiero, "GHISLIERI, Federico Fabio", *Dizionario Biografico degli Italiani*, 2000, Vol. 54.

5 Roberto Amerighi, *The "Presa di Bona" in Algeria. A feat of the Knights of St. Stephen in the 17th century*.Rome, Minimegaprint for ARLibri, 2017.

1544, Nov. 3rd – Giuliano Cesarini asks Cosimo I de' Medici to send "Cecco di Sandro", probably Altoni, to tutor Cesare Cesarini in fencing, in preparation for a duel with a Roman.[1]

> *[...] My lord the duke [Cosimo I] having been asked by lord Giuliano Cesarino to arrange for Cecco di Sandro [Francesco di Sandro Altoni] to go to serve lord Cesareo Cesarino and train him in fencing, because of a duel that he thinks will be between the aforesaid lord Cesareo and another Roman gentleman. He's pleased, as far as it belongs to his excellency, that the said Cecco go to serve him. It's true that his excellency wants this gentleman, servant of lord Giuliano, to come to an agreement with him about his wages. Your lordship will be happy to send for the said Cecco, and urge him to go and serve this lord, and take the servants of the Ceserina household, which he must hold in high regard for himself and his children. In the rest, this gentleman will come to an agreement with him, and it will remain between them how, when, and where he must go.*[2]

1553, Mar. 15th – Camillo Agrippa, an engineer from Milan, publishes a fencing book with copperplate illustrations, which is printed in Rome by Antonio Blado. It is dedicated to Cosimo I de' Medici.[3]

c.1560 – Francesco Ghislieri's son, Federico Fabio Ghislieri, is born in Bosco Marengo in Piedmont.[4]

1566 – Antonio Ghislieri becomes Pope Pius V. He appoints Francesco Ghislieri, a distant relative, as governor of the Papal fief of Ascoli.[5]

1564, Feb. 1st – Master Bastiano da Leccio is listed as fencing master in the Medici ducal household.[6]

---

1 Archivio di Stato di Firenze, *Mediceo del Principato*, Vol. 1171, f. 285r.

2 *Il duca mio signore [Cosimo I] essendo stato ricercato dal signor Giuliano Cesarino di operare che Cecco di Sandro [Francesco di Sandro Altoni] vadi a servire il signor Cesareo Cesarino et exercitarlo nella schirma [scherma], per cagione d'un duello che pensa dover essere tra il prefato signor Cesareo et un'altro gentilhomo romano. E' contento per quanto appartiene a sua excellentia che detto Cecco lo vadi a servire. E' ben vero che sua excellentia vuole che questo gentilhomo servitore del signor Giuliano s'intenda con lui della mercede sua. La signoria vostra sarà contenta di mandare per detto Cecco, et exhortarlo a ire a servire questo signore, et pigliare la servitù di casa Ceserina, della quale ha da tenere gran conto per sé et per li suoi figlioli. Nel resto questo gentilhomo converrà con lui, et resterà secco come, quando et dove doverrà andare.*

3 Agrippa, Camillo, *Trattato Di Scientia d'Arme, con un Dialogo di Filosofia*. Rome, 1553, e.g: https://onb.digital/result/107BDB08
There is some speculation that there might have been an earlier edition in 1527.

4 Brunelli, Giampiero, "GHISLIERI, Federico Fabio", *Dizionario Biografico degli Italiani*, 2000, Vol. 54.

5 *Ibid.*

6 Archivio di Stato di Firenze, *Mediceo del Principato*, Vol. 616, f. 262r. Leccio is a small village near Florence.

1567, Mar. 1st	– Antonio da Bologna is listed as fencing master of the pages in the ducal household.[1]
1567, May 25th	– Battista Viggiani, from Bologna, creates a fencing treatise for the Holy Roman Emperor Maximillian II.[2] It's based on a work by his deceased brother, Angelo, in 1551. Angelo had been a captain in the Imperial army c. 1530.[3]
1569	– Franscesco Ghislieri dies while campaigning against French Huguenots for the Papacy.[4]
1569, Aug. 27th	– Pope Pius V grants a bull elevating Cosimo I dei' Medici from Duke of Florence to Grand Duke of Tuscany. Altoni's fencing book is dedicated to Cosimo I, but uses the old title, so was probably written before this date.[5]
1571, May 16th	– A duel takes place in Prati, Rome between Giampaolo Baglioni and Antonio Salviati, over harsh words said in Florence a few days before. The matter couldn't be resolved despite the intervention of two lords. They exchanged half a dozen blows without injury, and then stopped the duel and made peace. Afterwards they were hosted in Rome by Ferdinando I de' Medici.[6]
1572	– Giovanni dall'Agocchie, from Bologna, publishes a fencing book in Venice.[7]

---

1 Archivio di Stato di Firenze, *Mediceo del Principato,* Vol. 616, f. 273r. He's also noted in letters on f. 448r & f. 298r.

Perhaps this was Antonino di Bologna who was an active officer in the Knights Hospitaller in Malta. Coming from a noble family in Palermo, Sicily, he was recorded at the siege of Malta (1565), and in the Order. See Lauriel, L.P., *Nuove Effermeridi Siciliane.* 1878, Vol VII, p. 185 & Aleccio, Matteo Perez d', *Siege of Malta*, 1582.

https://arbor.medialibrary.it/media/schedaopen?id=1507394

2 Viggiani del Montone, Angelo, *Trattato d'uno schermo di spada*. Bologna, 1567. Cod. 10723 HAN MAG. http://data.onb.ac.at/rec/AC13947205

3 https://theartofarms.substack.com/p/angelo-viggiani-del-montone

Citing Giordani, Gaetano, *Della venuta e dimora in Bologna del sommo Pontifice Clemente VII. per la coronazione di Carlo V. Imperatore: celebrata l'anno 1800 cronaca.* Alla Volpe, 1842.

4 Brunelli, Giampiero, "GHISLIERI, Federico Fabio", *Dizionario Biografico degli Italiani*, 2000, Vol. 54.

5 Acidini, Cristina, ed., *The Medici, Michelangelo, & the Art of Late Renaissance Florence.* Yale University Press, 2002, p. 309.

6 Archivio di Stato di Firenze, *Mediceo del Principato,* Vol. 616, f. 422v.

7 Agocchi, Giovanni dall', *Dell'arte di scrimia libri tre*. Venice, 1572, e.g: https://onb.digital//result/10B42FCA

1573, Jan. 23rd – Henry de Sainct Didier, a career soldier who fought in Piedmont, publishes a fencing book with woodblock illustrations, which is printed in Paris by Jean Mettayer and Matthurin Challenge. He recieves a royal copyright for ten years. He shows a familiarity with Neapolitan fencing jargon and has a section linking fencing practice to playing tennis.[1]

1574, Apr. 21st – Cosimo I de' Medici dies. Francesco I de' Medici succeeds him as Grand Duke of Tuscany.[2]

1574 – Francesco Ramires is listed as fencing master in the Ducal household, both before and after the change of Grand Duke.[3]

1575 – Angelo Viggiani's fencing text is printed in Venice by Giorgio Angelieri, with copperplate illustrations.[4]

1576 – Silvio Piccolomini is listed as part of the Ducal household.[5]

1578 – Alessandro Farnese, Prince of Parma is appointed governor of the Spanish Netherlands.[6]

1579 – Antonio da Bologna is back again as fencing master at the Medici court.[7]

1580, Oct. 16th – Michel de Montaigne, a famous french writer, passes through Augsburg on his way to Italy. He notes visiting a fencing hall and seeing longsword and messer fencing, comparing it to seeing a puppet show.[8]

---

1 Sainct Didier, Henry de, *Les secrets du premier livre sur l'espée seule*. Paris, 1573 e.g: https://gallica.bnf.fr/ark:/12148/bpt6k1520818c

2 https://www.treccani.it/enciclopedia/francesco-i-de-medici-granduca-di-toscana/

3 Archivio di Stato di Firenze, *Mediceo del Principato*, Vol. 616, f. 305 & f. 343r.

4 Viggiani dal Montone, Angelo, *Lo Schermo*. Venice, 1575. e.g: https://www.digitale-sammlungen.de/en/view/bsb10159894

5 Archivio di Stato di Firenze, *Mediceo del Principato*, Vol. 616, f. 382r.

6 Marek y Villarino de Brugge, André, *Alessandro Farnese: Prince of Parma: Governor-General of the Netherlands (1545-1592): v. II*. Los Angeles: MJV Enterprises, 2020, Vol. II, pp. 322-346.

7 Archivio di Stato di Firenze, *Mediceo del Principato*, Vol. 616, f. 367r.

8 Montaigne, Michel de, *The journal of Montaigne's travels in Italy by way of Switzerland and Germany in 1580 and 1581, Volume 1 (of 3)*. trans. Waters, 1903, Vol. 1, p. 128. https://www.gutenberg.org/cache/epub/70838/pg70838-images.html#i_fp128

1580, Nov. 18th – Michel de Montaigne visits Bologna, being hosted by the retired officer Blaise de Montluc, and comments on a Venetian fencing instructor and his French student:[1]

> *On the Friday we witnessed some sword play by a certain Venetian, who boasted of having discovered some new tricks in the art of fencing, all the others following his lead; in sooth his style differs vastly from the ordinary sword exercise. The best of his pupils was a young Bordelais named Binet.*

1580–81 – Federico Ghislieri serves as an officer in the Tuscan *"Prospero Colonna"* tercio (a unit of military administration in Spanish service), campaigning for the Spanish in Portugal.[2]

1581, Jun. 26th – Silvio Piccolomini hosts Michel de Montaigne for dinner in Florence. Michael records some of the conversation:[3]

> *On Monday I dined with Signor Silvio Piccolomini, a man of well-known worth and especially skilled in the art of fence. Many subjects came up for discussion, as there was a large gathering of other gentlemen.*
>
> *He has a poor opinion of the art of fencing as practised by the Italian masters, such as Veniziana di Bologna, Patinostraro, and others. Indeed, the only one he favoured was a certain pupil of his own, who now lives at Brescia, where he teaches the art to the gentle–folk of the city. He declared that the ordinary teachers of fencing follow neither art nor rule, and he condemned especially the far-reaching lunge which leaves your sword at the mercy of your adversary, and the practice of making one assault after another and then coming to a pause. This, he declared, was entirely against all experience of sword-play. He was on the eve of printing a treatise on this subject.*
>
> *In speaking of warfare he made light of artillery, greatly to my satisfaction, and he praised Machiavelli's work,* Della Guerra, *and was quite in agreement with the writer's opinions. He declared that the most skilful military engineer living was one in the service of the Grand Duke in Florence.*

Paternostro is mentioned in other sources[4] and a fencing text attributed to him survives.[5] The Venetian in Bologna could be the one referenced by

1 Montaigne, Michel de, *The journal of Montaigne's travels in Italy by way of Switzerland and Germany in 1580 and 1581, Volume 2 (of 3)*. trans. Waters, 1903, Vol. 2 p. 38.
https://www.gutenberg.org/cache/epub/70839/pg70839-images.html#i_fp038

2 Brunelli, Giampiero, "GHISLIERI, Federico Fabio", *Dizionario Biografico degli Italiani*, 2000, Vol. 54.

3 Montaigne, Michel de, *The journal of Montaigne's travels in Italy by way of Switzerland and Germany in 1580 and 1581, Volume 3 (of 3)*. trans. Waters, 1903, Vol. 3 pp. 103–4.
The dating is based on the Feast of St John being Midsummer's Day, June 24th.
https://www.gutenberg.org/cache/epub/70840/pg70840-images.html

4 D'Alessandri, Torquato. *Il cavaliere compito*. Viterbo, 1609. p.109.

5 Cavalcabo , Girolamo, *Traité ou instruction pour tirer des armes*. Paris, 1595.

Michel de Montaigne. The pupil in Brescia has sometimes been conjectured as Fallopia.[1]

1581–84 – Federico Ghislieri joined Alessandro Farnese's army in Flanders, serving as a captain with a roving commission.[2]

1582, Apr. 1st – Jerónimo Sánchez de Carranza, a Spanish courtier, publishes a fencing book.[3]

1584 – Federico Ghislieri serves in the Cologne War, as part of Alessandro Farnese' forces brought south from Flanders to assist Catholic interests. He participates in the Siege of Bonn, in the cavalry of Ernest of Bavaria.[4]

1584 – Alfonso Falloppia, ensign of Bergamo fortress, has a fencing book published there by Comin Ventura. It is dedicated to Ranuccio Farnese, the adolescent son of Alessandro.[5]

c.1585(?) – Federico Ghislieri composes the first manuscript copy of his fencing book.[6]

1586, Oct. 30th – Silvio Piccolomini writes to Francesco I, providing news of Italian soldiers campaigning in Flanders under Alessandro Farnese. He describes Irish runners armed with bows forming part of the English army. The letter is signed by Silvio himself.[7]

1 https://wiktenauer.com/wiki/Alfonso_Falloppia
2 Brunelli, Giampiero, "GHISLIERI, Federico Fabio", *Dizionario Biografico degli Italiani*, 2000, Vol. 54.
3 Carranza, Jerónimo Sánchez de, *De la Filosofia de las Armas.* Sanlúcar de Barrameda, 1582. The work was completed in 1569, but not published until later. e.g: https://gallica.bnf.fr/ark:/12148/bpt6k8574246
4 Brunelli, Giampiero, "GHISLIERI, Federico Fabio", *Dizionario Biografico degli Italiani*, 2000, Vol. 54.
5 Falloppia, di Alfonso, *Nuovo et brieve modo di schermire*. Bergamo: Comin Ventura, 1584 e.g: https://catalogue.bnf.fr/ark:/12148/cb30417239j
6 A manuscript copy of Ghislieri's book exists, but I haven't been able to verify its dating. See: Gotti, Roberto (2023). "The Dynamic Sphere: Thesis on the Third State of the Vitruvian Man." *Martial Culture and Historical Martial Arts in Europe and Asia: 93–147.* Ed. by Daniel Jaquet; Hing Chao; Loretta Kim. Springer.
7 Archivio di Stato di Firenze, *Mediceo del Principato*, Vol. 4225, ff. 71r-73r.

1587	– Silvio Piccolomini is listed as the Ducal household.[1]
1587, Apr. 22nd	– Federico Ghislieri's fencing book is printed by Erasmo Viotti in Parma. The text twice mentions that Ghislieri learnt fencing from Silvio Piccolomini.[2] It's printed without engraved illustrations. Some copies have delightful pen and wash illustrations.[3] One copy has a crude illustration, perhaps reminiscent of the manuscript copy, as well as hand written corrections and perhaps printer's notes.[4] There are two different dedications printed. Some copies are dedicated to Ranuccio Farnese, the son of Alessandro. Others are dedicated to Antonio Pio Bonelli, a future soldier, later the Marquis of Cassano.[5]
1587, Oct. 3rd–7th	– Silvio Piccolomini meets Francesco I who's travelled to Brussels.[6]
1587, Oct. 20th	– Francesco I de' Medici dies of malaria in Italy. His brother, Ferdinando I de' Medici, succeeds him as Grand Duke of Tuscany.[7]
1588	– Silvio Piccolomini is part of the household of Giovanni de' Medici, in Flanders. Giovanni was a military officer, and illegitimate son of Cosimo I, in Flanders from 1587–89. His staff write to the Tuscan court, asking for financial aid.[8]
1588, July 26th	– Angelo Viggiani's fencing text is republished in Bologna, with a new introduction by Zacharia Cavalcabo.[9]
1588, Mar. 12th	– Silvio Piccolomini is mentioned as staying with Giovanni di Cosimo de' Medici. [10]
1588, Oct. 7th	– Silvio Piccolomini is mentioned again as staying with Giovanni di

1 Archivio di Stato di Firenze, *Mediceo del Principato*, Vol. 616, f. 391r.

2 Ghislieri describes himself as the creation of Sir Silvio Piccolomini on p. iii of the Farnese introduction, and describes the ideal fencing posture as discovered by Silvio Piccolomini on p. 94, in the analogy to siegecraft.

3 The most accessible illustrated copy is in the Bibliothèque nationale de France, which has the Ranuccio Farnese introduction. NUMM–1518476.
https://gallica.bnf.fr/ark:/12148/bpt6k1518476n/

4 Folger Shakespeare Library, SU101 .G5 1587 Cage.
https://digitalcollections.folger.edu/bib75529-77562

5 The British Library copy is dedicated to Antonio Pio Bonelli, and is not illustrated. It doesn't mention Piccolomini in its dedication, only in the text on p. 94.

6 Archivio di Stato di Firenze, *Mediceo del Principato*, Vol. 5151, p. 215r.

7 di Gino, Benzoni, "FRANCESCO I de' Medici, granduca di Toscana", *Dizionario Biografico degli Italiani*, 1997, Vol. 49.

8 Archivio di Stato di Firenze, *Mediceo del Principato*, Vol. 5151, f. 91r.

9 Viggiani dal Monte, Angelo, *Trattato dello schermo d'Angelo Vizani*, ed. Zacharia Cavalcabo, Bologna, 1588, e.g:
http://data.onb.ac.at/rep/10AC83BA

10 Archivio di Stato di Firenze, *Mediceo del Principato*, Vol. 5151, f. 69r. &
https://www.firefiandre.be/artikels/was-er-meer-dan-n-giovanni-de-medici-in-de-lage-landen

	Cosimo de' Medici.[1]
1590	– Federico Ghislieri raises two thousand infantry to suppress disorder in the Papal States during a succession crisis.[2]
1591	– Federico Ghislieri raises a company of infantry from the Papal States, to join the Catholic League in the French Wars of Religion.[3]
1592	– Silvio Piccolomini joins the Knights of St. Stephen, a Tuscan military order based out of Pisa. In time, he became the Grand Constable of the Order.[4]
1592, Apr. 25th	– Silvio Piccolomini meets the entourage of Cardinal Francesco Sforza, travelling from Rome to Florence for the baptism of Cosimo II.[5]
1592, Aug. 15th	– Silvio Piccolomini again meets the entourage of Cardinal Francesco Sforza, as they travel to meet Duke Ferdinando I.[6]
1593, Dec. 20th	– Silvio Piccolomini is listed as Captain of the Pisan cavalry in a list of Tuscan officers.[7]
1595	– Silvio Piccolomini is sent on the Hungarian campaign as a Captain of Rudolph II's infantry.[8]
1595	– Girolamo Cavalcabo, from Bologna, publishes a fencing book in Paris. Although published in French, an original Italian manuscript survives. It also contains a section attributed to the late Paternoster of Rome.[9]
1595, Aug. 7th	– Eleanora de' Medici-Gonzaga, Duchess of Mantua, writes to Ferdinando I, informing her uncle of arrangements made for Silvio Piccolomini's travel on the way to Hungary.[10]
1595, Sep. 21st	– Federico Ghislieri, as sergeant major and infantry commander of a Papal Tercio in Hungary, commands artillery at the Siege of Visegrád, forcing

1 Archivio di Stato di Firenze, Mediceo del Principato, Vol. 5151, f. 209r.

2 Brunelli, Giampiero, "GHISLIERI, Federico Fabio", *Dizionario Biografico degli Italiani*, 2000, Vol. 54.

3 *Ibid.*

4 Guarnieri, Gino, *The Order of Santo Stefano, in his internal organization. Lists of knights belonging to the order with chronological references, of homeland, of title, of dressing of Dresses: 1562–1859*, Pisa, Giardini Ed., 2012. p. 167.

5 Archivio di Stato di Firenze, *Guardaroba Medicea Diari di Etichetta*, Vol. 3, p. 26.

6 Ibid. p. 60.

7 Archivio di Stato di Firenze, *Mediceo del Principato*, Vol. 616, f. 419r.

8 Spreti, Vittorio, *Encyclopedia Storico Nobiliare Italiana*, Milan, Encyclopedia Storico Nobiliare Italiana, 1928–1935. Vol. V, p. 330.

9 Cavalcabo, Girolamo, *Traité ou instruction pour tirer des armes*. Paris, 1695, e.g: https://bibnum.institutdefrance.fr/ark:/61562/mz17837
The manuscript is Bibliothèque nationale de France. Département des manuscrits. italien 1527.

10 Archivio di Stato di Firenze, *Mediceo del Principato*, Vol. 2942.

	the Ottoman garrison to surrender.[1]
1597, Aug. 21st	– Federico Ghislieri serves in the same roles, in a Papal Tercio commanded by Mario I Farnese, Duke of Latera. Once in the field, he was appointed as a general sergeant, with a salary of 140 scudi a month, by the expedition commander, Gian Francesco Aldobrandini. He co–commands the Papal troops in capturing Papa.[2]
1597, Nov.	– Federico Ghislieri continues to lead Papal forces fighting at Vác, following the Catholic failures at Győr and Komárom.[3]
1597–98	– The University of Cambridge brings three cases against individuals for giving fencing lessons, a contrast with Italian education.[4]
1598, Apr. 4th	– Federico Ghislieri writes an account of the siege of Györ, a Hungarian town occupied by the Ottomans, for Emperor Rudolf II.[5]
c. 1600	– Camillo Palladini, from Bologna, creates an illustrated fencing manuscript.[6]
1600, Nov. 12th	– Ottavio Piccolomini, son of Silvio, is born in Pisa. He went on to have a military career through the Thirty Years War.[7]
1601, May 17th	– Marco Docciolini published a fencing book in Florence, dedicated to Giovanni de' Medici.[8] At this time Giovanni was campaigning in Hungary against the Ottomans, but had been a Tuscan diplomat for a few years.[9]
1601, Jul. 23rd	– Federico Ghislieri is mentioned as a sergeant major of Papal forces for a Hungary campaign to aid Duke Ferdinand of Styria in taking Nagykanizsa.[10]

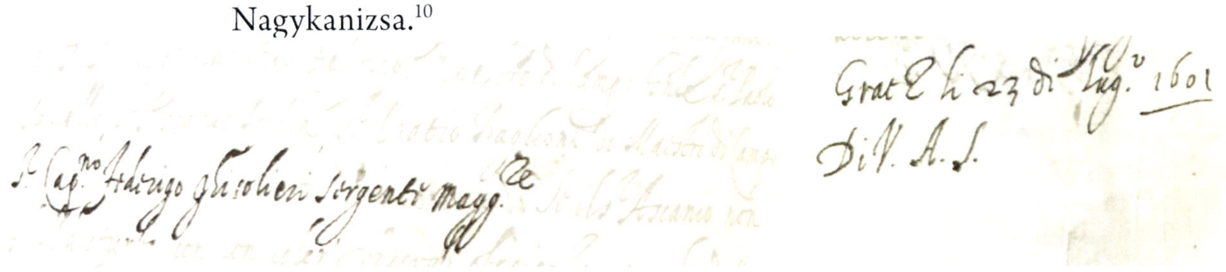

1 Brunelli, Giampiero, "GHISLIERI, Federico Fabio", *Dizionario Biografico degli Italiani*, 2000, Vol. 54.
2 *Ibid.*
3 *Ibid.*
4 https://archivesearch.lib.cam.ac.uk/repositories/2/archival_objects/54890
5 Ms.: National Central Library of Florence; ed. Maggiorotti – Banfi 1933. pp. 44–45.
6 Palladini, Camillo, Howard de Walden Library MS 14/10
7 https://www.treccani.it/enciclopedia/piccolomini-ottavio-duca-d-amalfi-e-principe-dell-impero_(Enciclopedia-Italiana)/
8 Marco Docciolini, *Trattato in Materia di Scherma*, Parma, 1601. e.g: BSB-ID 991033848699707356
9 https://www.treccani.it/enciclopedia/giovanni-de-medici_(Dizionario-Biografico)/
10 Archivio di Stato di Firenze, *Mediceo del Principato*, Vol. 5155, p. 27r.

1601, Sep. – Federico Ghislieri is promoted to Sergeant Major General, following Aldobrandini's death. He's now second in command of Papal forces, and paid 300 scudi a month.[1]

1601, Sep. 30th – Salvator Fabris, from Padua, wrote a fencing manuscript for John Frederick of Holstein-Gottorp, a Prince-Archbishop of Bremen, Lübeck and Verden.[2]

1601, Oct. 1st – Silvio Piccolomini and his wife escort the French Ambassador, who lands in Livorno en route to Rome. He's recorded as Captain of the Pisan cavalry.[3]

1601, Oct. 28th – Catholic forces lead an unsuccessful assault on Nagykanizsa, after just under a month of camping in a march besieging the Ottomans. Federico Ghislieri's diligence leading the siege is noted.[4]

1601, Nov. 17th – Rumour of an Ottoman relief force sparks a panicked retreat from Federico Ghislieri's forces at Nagykanizsa. Ghislieri falls ill on the march, and while convalescing writes an explanation of his behaviour:[5]

> *Ghislieri also reported that during the two-month siege, he was repeatedly struck by the wind of death, two musket bullets got stuck in his clothes, and several times people standing nearby were hit by cannon fire. During the last attack, on St. Simon's Day, a fragment of a cannonball hitting a tree next to Ghislieri wounded him so severely that during the shameful retreat following the unsuccessful siege, he was only able to be evacuated to Pettau [today: Ptuj, Slovenia], where he was forced to spend a long time recovering in a hospital set up for him, along with the other wounded who couldn't be transported.*[6]

---

1 Brunelli, Giampiero, "GHISLIERI, Federico Fabio", *Dizionario Biografico degli Italiani*, 2000, Vol. 54.

2 Fabris, Salvatore, *Scientia e Prattica dell'Arme*, 1601, GI kongelig Samling 1868 4040

3 Archivio di Stato di Firenze, Guardaroba Medicea Diari di Etichetta, Vol. 3, p. 168.

4 Brunelli, Giampiero, "GHISLIERI, Federico Fabio", *Dizionario Biografico degli Italiani*, 2000, Vol. 54.

5 *Ibid.*

6 Vatican Apostolic Archives, Fondo Borghese III, vol. 93 B2, fol. 43v. See:
Szovák Márton: 'Federico Ghislieri hadmérnök szakírói tevékenysége különös tekintettel a tizenöt éves háborúra', *HADTÖRTÉNELMI KÖZLEMÉNYEK, Vol. 134*: (1), 2021: pp. 73-101.

1602, July 31st	– Jean Héroard, doctor of King Henry IV of France, distracts the injured King with some play-fencing:1
	July 31st [1602], Wednesday. — Impatient to go out; he meets the King; taken in a carriage into the forest to see the deer chased by the King, who had dined at Forqueil, where the assembly had taken place. Carried to the King, in his bed, injured from a fall, chasing the deer. He holds a baton; I take a piece of stick, I strike it against his baton to fence; the game pleases him, he chases me laughing throughout the room. The rest of the day is peaceful and very cheerful. — This day, at five o'clock, Marshal Biron had his head cut off at the Bastille.
1603, Jul. 10th	– Silvio Piccolomini and his wife visit the household of Pompeo del Monte Santa Maria, a Tuscan noble.[2]
1603, Aug. 1st	– Giovanni Alberto Cassani, an officer in the Spanish armies from Piedmont, publishes a military treatise in Naples. Although mostly concerned with tactics, it gives a few pages on swordplay, showing a simple system based on combinations of cuts.[3]
1603, Nov. 6th	– Silvio Piccolomini hosts a visit from Count Giulio Pepoli in Florence.[4]
1604	– Federico Ghislieri pursues various diplomatic postings for the Papacy.[5]
1604, Jan. 28th	– James I & IV of England and Scotland recieves a fencing master for his children from the French ambassador.[6] The fencing master is Charles Guérolt, who is mentioned again in court accounts in 1604 & 1606.[7]
1604, Feb. 22–25th	– Silvio Piccolomini is mentioned in relation to a visit to Florence by Duke Camillo Conti di Carpineto.[8]
1604, Aug. 4th	– Silvio Piccolomini lodges in Tuscany.[9]

1 Jean Héroard, *Journal de Jean Héroard sur l'enfance et la jeunesse de Louix XIII (1601-1628)*, 1868, Paris. p. 32.

2 Archivio di Stato di Firenze, Guardaroba Medicea Diari di Etichetta, Vol. 3, p. 194

3 Cassani, Giovanni Alberto. *Essercitio militare, il quale dispone l'huomo à vera cognitione del scrimire di spada, e dell'ordinare l'essercito à battaglia etc.*, Naples, 1603 e.g:
https://books.google.co.nz/books?id=nBtaAAAAcAAJ

4 Archivio di Stato di Firenze, Guardaroba Medicea Diari di Etichetta, Vol. 3, p. 197.

5 Brunelli, Giampiero, "GHISLIERI, Federico Fabio", *Dizionario Biografico degli Italiani*, 2000, Vol. 54.

6 *'Venice: January 1604', in Calendar of State Papers Relating To English Affairs in the Archives of Venice, Volume 10, 1603-1607,* ed. Horatio F Brown (London, 1900), British History Online
https://www.british-history.ac.uk/cal-state-papers/venice/vol10/pp126-130 [accessed 11 September 2025].

7 Devon, F. *Issues of the Exchequer: Being Payments Made Out of His Majesty's Revenue During the Reign of James I,* Great Britain Record Commission (1836) p. 17 & p. 34.

8 Archivio di Stato di Firenze, Guardaroba Medicea Diari di Etichetta, Vol. 3, p. 205.

9 Archivio di Stato di Firenze, Guardaroba Medicea Diari di Etichetta, Vol. 3, p. 219.

1604, Sep. 1st	– Silvio Piccolomini again lodges in Tuscany.[1]
1605	– Guido Biandrate di San Giorgio was recorded as Maestro di Campo of a Milanese tercio in Flanders for the Hapsburgs, but due to increasing tensions between the Spanish Hapsburgs and the French leaning Savoy, left for Charles Emmanuel's army in 1607.[2]
1605, May 3rd	– Federico Ghislieri is appointed Sergeant Major General of the Papal States following the death of Clement VIII in May, however he leads the Knights of St. Stephen in an amphibious attack on Preveza, having been appointed Maître de Camp of the General Infantry by Ferdinando I de' Medici.[3] The raid is successful, and Federico wrote to Pope Paul V telling him that he'd planted the banners on a rock "in the village of the Turks as if we'd been in Christendom."[4] A pamphlet is printed telling of the raid, which provides a lot of detail for Ghislieri lead the infantry forces.[5] Afterwards he returns to Papal service.
1606, Sep. 25th	– Salvator Fabris publishes a fencing book in Copenhagen, dedicated to the Chrstian III of Denmark.[6]
1606, Feb. 10th	– Nicoletto Giganti writes a fencing book in Venice, dedicated an adolescent Cosimo II de' Medici, who became Grand Duke of Tuscany three years later.[7]
1606, May	– Federico Ghislieri aids Mario Farnese, the lieutenant general of Papal forces, inspecting coastal fortifications as tensions rose with the Venetian Republic.[8]
1606, Jul.	– Federico Ghislieri raises 8000 infantry in Ancona, Marche, in preparation for potential war with Venice.[9]

---

1 Archivio di Stato di Firenze, Guardaroba Medicea Diari di Etichetta, Vol. 3, p. 221.

2 He led the Savoyard infantry during the First War of Monferrato. See Maurizio Arfaioli, *Bastion of Empire: The Italian terzo vecchio of the Army of Flanders (1597-1682)*, Jan. 2022, The Journal of Military History 85(1), pp. 27-50.

3 Brunelli, Giampiero, "GHISLIERI, Federico Fabio", *Dizionario Biografico degli Italiani*, 2000, Vol. 54 & Pujeau, E. (2015) 'How to take a Fortress? The Wrong and the Right Way: Preveza 1538 and 1605.', *Πρεβεζάνικα Χρονικά*, 0(51–52), pp. 119-134. doi: 10.12681/prch.28292

4 Arch. segr. Vatican, Borghese Fund, III, 11a–b, c. 47v.

5 *La Relazione dell'impresa della Prevesa fatta per ordine del Serenissimo Gran Duca di Toscana. Dalle Galere della Religione di Santo Stefano seguita a 3 di Maggio 1605, Florence 1605*, e.g: https://books.google.co.nz/books?id=vgBiT_bPOuYC

6 Fabris, Salvator, *De lo Schermo, overo Scienza d'Arme*. Copenhagen, 1606, e.g: https://gallica.bnf.fr/ark:/12148/bpt6k1507286r

7 Nicoletto Giganti, *Scola, overo teatro*. Venice, 1606.

8 Brunelli, Giampiero, "GHISLIERI, Federico Fabio", *Dizionario Biografico degli Italiani*, 2000, Vol. 54.

9 ibid.

1607	– Guido Biandrate di San Giorgio leaves Flanders for Charles Emmanuel's service due to increasing tensions between the Spanish Hapsburgs and the French leaning Savoy. He led the Savoyard infantry during the the War of Montferrat Succession.[1]
1607, Oct.	– Venice and the Papacy reach a détente, Federico Ghislieri is appointed to Pope Paul V's military congregation but is out of favour for his bellicosity.[2]
1607, Sep. 15th	– Silvio Piccolomini commands an expedition by the Knights of St. Stephen, which sacks Bona in Algeria.[3] A pamphlet is also printed.[4]
1608, Jan. 16th	– Julius Caesar (the younger), the son of an English Privy Councellor of the same name, is killed in Padua in a dispute over behaviour at Tagliaferro's fencing school. He had been struck on the left hand and required medical treatment, then sought revenge. The death led to a minor diplomatic incident that took months to resolve.[5]
1608, May 25th	– Nicoletto Giganti writes a second fencing book, in Pisa, this time it's dedicated to a Venetian knight Christofano Chigi, a member of the same Tuscan learned society as Piccolomini. Now Giganti describes himself as master at arms for the Knights of St. Stephen.[6]
1608, Oct.	– Federico Ghislieri is appointed Sergeant General of Papal forces, with an annual salary of 1200 scudi.[7]
1609, Feb. 7th	– Ferdinando I de' Medici dies. He's succeeded by his son Cosimo II de' Medici.[8]

1 Maurizio Arfaioli, 'Bastion of Empire: The Italian terzo vecchio of the Army of Flanders (1597-1682)', *The Journal of Military History*. Jan. 2022, 85(1): pp. 27-50.

2 Brunelli, Giampiero, "GHISLIERI, Federico Fabio", *Dizionario Biografico degli Italiani*, 2000, Vol. 54.

3 Archivio di Stato di Firenze, Mediceo del Principato, Vol. 5052, f. 483r, & Vol.1305, f. 295r.
See also Roberto Amerighi, *La "Presa di Bona" in Algeria. Un'impresa dei Cavalieri di Santo Stefano nel XVII secolo*, Roma, Minimegaprint per ARLibri, 2017.

4 *Relazione del viaggio, e della presa della città di Bona in Barberia. Fatta per commessione del Sereniss. Gran Duca di Toscana in nome del Sereniss. Prencipe suo Primogenio, dalle Galere della Religione di Santo Stefano. Il dì 16 di Settembre 1607. Sotto il Comando di Siluio iccolomini Gran Contestabile di detta Religione, & Aio del medesimo Prencipe*, Rome 1607 e.g:
https://books.google.co.nz/books?id=nlFFAAAAYAAJ

5 Venice: October 1608', in *Calendar of State Papers Relating To English Affairs in the Archives of Venice*, Volume 11, 1607–1610, ed. Horatio F Brown (London, 1904), British History Online
https://www.british-history.ac.uk/cal-state-papers/venice/vol11/pp174-185 [accessed 11 September 2025].

6 Nicoletto Giganti, *Libro secondo*. Pisa, 1608. The only known copy is at the Howard de Walden library HDW GIG 1608

7 Brunelli, Giampiero, "GHISLIERI, Federico Fabio", *Dizionario Biografico degli Italiani*, 2000, Vol. 54.

8 https://www.treccani.it/enciclopedia/ferdinando-i-de-medici-granduca-di-toscana_(Enciclopedia-Italiana)/

1609, Feb. 20th – The Dauphin of France, a seven year old boy, breaks his fencing sword trying to surprise his teacher, Girolamo Cavalcabo with a forehand blow.[1]

1609, May 5th – Torquato d'Alessandri publishes a fencing book in Viterbo, the Papal States. On p. 107, he gives a list of fencing instructors:[2]

Rome	**Bologna**
Oratio & Cesare Cavalcabo	*Francesco & Vicenzo Marcelli (known as the Abruzzesi)*
Camillo Palladini	*Appio Castelli*
	Giovanni Angelo Paternostraro
	Antonio Rinaldi

1609 – A fencer from Verona identified as "Joann Guerini" petitioned the Stasbourg civic council to teach fencing there over the winter. Despite claiming to have met royalty, his request was denied.[3]

Joann Guerini of Verona, a fencer, submits a supplication. He requests to allow him to practice his knightly art here this winter, and to inform young gentlemen and [those] of the nobility. He wishes to behave properly. He beats the drum, requests even more since he also met a queen[?], thus even more reason to allow it.

Recognized. His request has been denied. Herr Wend[?], Herr Eg[?]

Margin: Fencer, Italian, for residence this winter.[4]

1610 – Federico Ghislieri writes regarding military command. He holds a number of Roman offices and receives various pensions for the next few years.[5]

1610, May 4th – The nine year old Dauphin, the future Louis XIII of France, receives corporal punishment for his behaviour the previous day. Not only had he thrown asparagus at a courtier, he also hit his fencing instructor Girolamo Cavalcabo in the eye while Cavalcabo attempted to correct his footwork.

---

1 Jean Héroard, *Journal de Jean Héroard sur l'enfance et la jeunesse de Louix XIII (1601-1628)*. 1868, Paris. p. 384.

2 Torquato d'Alessandri, *Il Cavalier Compito: Nel quale si discorre d'ogni scienza, di ragion di Stato, di Medicina, di Metheora, di dubbi Caualiereschi*. Viterbo, Italy, 1609. e.g: https://books.google.co.nz/books?id=AypUAAAAcAAJ

3 Thanks to Liam Clark for bringing this to my attention. https://archives.strasbourg.eu/ark:/39332/vta0552654857f427d7/daogrp/0/259

4 *Joann Guerini von Verona ein fechter vber[=] | gibt suppl[icati]on bitt Jme alhie dissenn [= diesen] winter | seine Ritterliche Kunst exerciren zulassen | vnd Junge h[err]n vndvon Adel zu informiere[n] | Will sich aller gepuer verhalten. Trom[m]et[?] | bitt noch mehr d[erwei]l ert[?] auch ein konginn[?] er[=] | fahren desto mehr zugonne[n] Erk[annt] Jst sein | begeren abgeshlagenn. H Wend H Eg*

5 BBrunelli, Giampiero, "GHISLIERI, Federico Fabio", *Dizionario Biografico degli Italiani*, 2000, Vol. 54 & Bibl. apost. Vaticana, Barb. lat. 4344, ff. 53r-54v.

	Louis' father is assassinated a few days later.[1]
1610, Apr. 8th	– Ridolfo Capoferro da Cagli, master of the German students at the University of Siena, writes a fencing book. In it, he praises Silvio Piccolomini.[2] The introduction of the book features the crest of the Accademia dei Filomati, a Tuscan learned society. Contemporary members of the Filomati include Silvio Piccolomini and Christofano Chigi, to whom Giganti dedicated his 1608 book.[3]
1610, Oct. 28th	André des Bordes (Abraham Racinot; 1582–1625) publishes a fencing book in Nancy, France.[4] It seems derivative of a fencing manuscript written by Camillo Palladini.[5] Palladini mentions the following masters:

Chapter IV – 'Old Masters'	**Chapter XVIII – 'Masters Seen'**
Tapa of Milan	*Sir Francesco fiorini 'the deaf'*
the Beccaroni	*Palela*
Mancino from Bologna[6]	*Gardano*
	Bandinelle

1611,	– John Florio, a member of the English royal household, publishes an Italian-English dictionary full of fencing terminology.[7]
1612	– Silvio Piccolomini dies in Florence.[8]

---

1 Jean Héroard, *Journal de Jean Héroard sur l'enfance et la jeunesse de Louix XIII (1601-1628)*. 1868, Paris. p. 433.

2 Ridolfo Capo Ferro da Cagli, *Gran Simulacro dell'Arte e dell'Uso della Scherma*. Siena, 1610 e.g: https://hdl.handle.net/2027/gri.ark:/13960/t8sc1ph47

3 L. Sbaragli, 'I tabelloni degli Intronati', *Bullettinosenesedi storia patria*, xx. 1942, pp. 198-206.

4 André des Bordes, *Discours de la théorie de la pratique et de l'excellence des armes*. Nancy, France, 1610. http://bmn-renaissance.nancy.fr/items/show/1229

5 Camillo Palladini, *Discorso sopra l'arte della scherma.* (MS 14.10), c. 1600.
The only known copy is in the Howard de Walden library, HDW PAL 16--

6 Usually identified with Giacomo Mancino da Bologna , a mercenary captain active 1493–d. 1508.

7 Florio, John, *Queen Anna's New World of Words, or Dictionarie of the Italian and English tongues. London,* 1611
https://www.pbm.com/~lindahl/florio/

8 Vittorio Spreti, Encyclopedia Storico Nobiliare Italiana, Milan, *Encyclopedia Storico Nobiliare Italiana, 1928-1935. Vol. V,* p. 330.

1616, Jan. 17th – Federico Ghislieri hosts Galileo at his Roman apartments for dinner, as mentioned in a letter by the Spanish ambassador:[1]

You'd be delighted to hear Galileo argue, as he often does, in the midst of some fifteen or twenty persons who attack him vigorously, now in one house, now in another. But he's so well buttressed that he laughs them off; and although the novelty of his opinion leaves people unpersuaded, yet he shows that most of the arguments with which his opponents try to overthrow him are spurious. Monday in particular, in the house of Federico Ghisilieri, he performed marvellous feats. What I liked most was that, before answering objections, he improved on them and added even better ones, so that, when he demolished them, his opponents looked all the more ridiculous.

1616, Oct. 8th – A letter from Florence to Cosimo I de' Medici describes a fight between a group of gentlemen and a fencing instructor:[2]

Ricasoli [Giulio] is entangled not only in that affair but also in all his other matters, because Bettino and his brother, the knight [Ottaviano], caused a stir on Tuesday evening that Her Most Serene Highness views most unfavourably. For the two of them, along with Sir Strozzi, Filippo Caponi, and Sir Rimbotti, came to confront Cherubino the fencer and his brother, the latter of whom was wounded, his right hand left almost touching the ground.[3]

1617 – Federico Ghislieri leaves Rome to join Charles Emmanuel I, Duke of Savoy, commanding forces during the War of Montferrat Succession, appointed Fieldmaster General and Lieutenant of the regiment of the Duke's guard by the end of the war that year. He fought at Crevacuore and San Damiano d'Asti.[4] Ghislieri then wrote a number of books on fortifications, most of which were lost in a fire at the Turin archives. The only one to survive was the *Discourse on Fortifications* from 1617.[5]

1618, Mar. 20th – Federico Ghislieri writes a book for Charles Emmanuel I on military strategy, *Military Speeches*.[6]

---

1 Antonio Querengo to Alessandro D'Este, 20 January 1616, *Opere*, vol. 12, pp. 226-227.

2 Archivio di Stato di Firenze, *Mediceo del Principato*, Vol. 5150, f. 89r.

3 *Il Ricasoli [Giulio] attraversa non solo quel negotio ma ancora tutti gli altri suoi affari poichè Bettino e il cavalierere suo fratello [Ottaviano] ne fecceno una martedì sera che Sua Altezza Serenissima la intende molto male poichè lor dua, e il cavaliere Strozzi e Filipo Caponi, e il cavalierr Rimbotti, si trovorno ad affrontare Cherubino schermitore e un suo fratello, il quale restò ferito e la man dritta poco meno che in terra*

4 Brunelli, Giampiero, "GHISLIERI, Federico Fabio", *Dizionario Biografico degli Italiani*, 2000, Vol. 54 & Pierpaolo Merlin, *Il Monferrato. Un territorio strategico per gli equilibri europei del Seicento*, in Pierpaolo Merlin e Frédéric Ieva (a cura di), *Monferrato 1613, La vigilia di una crisi europea*. Roma, Viella, 2016, pp. 15-30. ISBN 9788867286126.

5 Turin, *National Bibl.*, N.II.4.

6 Brunelli, Giampiero, "GHISLIERI, Federico Fabio", *Dizionario Biografico degli Italiani*, 2000, Vol. 54

1619	– Federico Ghislieri writes two books on fortifications for Charles Emmanuel I: '*Treatise in response to your highness the Duke of Savoy whether the city of Turin should be fortified...*' & '*Treatise that in conflict, the defender can't counter the battery of the attacker.*' [1]
1619	– Giovanni Battista Gaiani publishes a book on tournaments dedicated to a "Conte Guido San Giorgio", presumably Guido Biandrate San Giorgio Aldobrandini, a Savoyard military commander who served in Flanders.[2]
1619, Apr. 15th	– Giovanni Battista Gaiani, a career soldier, writes a fencing book in Genoa. He dedicates it to the two military sons of Charles Emmanuel I. It's notable that he shows an ignorance of previous texts on mounted fencing, which would have included Ghislieri's.[3]
1620	– Federico Ghislieri is made Marquis of Rocco by Charles Emmanuel I, allowing him to legitimise a daughter, Barbara.[4]
1621	– Jacopo Monesi is recorded as fencing master and one of several masters to the pages at the Tuscan court.[5]
1621, Feb. 28th	– Cosimo II de' Medici dies. He's succeeded by his ten year old son, Ferdinando II de' Medici, who rules until 1670.[6]
1621, Nov. 15th	– Bonaventura Pistofilo, from Ferrara, publishes a book in Siena on the use of the pike, halberd and musket. It's dedicated to Kenelm Digby, an English knight.[7]
1622(?)	– Federico Ghislieri dies in Turin, Savoy.[8]
1625, Jan. 28th	– Andre des Bordes is killed. At the time of his death, he owned a 1609 printing of Cavalcabo and Paternoster.[9]
1628, Sep. 15th	– Duke of Mantua Carlo I Gonzaga recommends Frenchman Stéfan Pélican, master of arms, to Grand Duke of Tuscany Ferdinando II.[10]

& Arch. of State of Turin, *Bibl. ancient*, Z.II.27.

1 Brunelli, Giampiero, "GHISLIERI, Federico Fabio", *Dizionario Biografico degli Italiani*, 2000, Vol. 54. & Arch. of State of Turin, *Bibl. ancient*, J.a.VII.23 & Arch. of State of Turin, *Bibl. ancient*, Z.II.32.

2 Maurizio Arfaioli, '*Bastion of Empire: The Italian terzo vecchio of the Army of Flanders (1597–1682)*', *The Journal of Military History*. Jan. 2022, 85(1), pp. 27-50.

3 Gaiani, *Arte di maneggiar la spada a piedi et a cavallo*, Loano, 1619. e.g: https://books.google.co.nz/books?id=949mAAAAcAAJ

4 Brunelli, Giampiero, "GHISLIERI, Federico Fabio", *Dizionario Biografico degli Italiani*, 2000, Vol. 54.

5 Archivio di Stato di Firenze, *Miscellanea Medicea*, 369, c.707.

6 https://www.treccani.it/enciclopedia/cosimo-ii_(Enciclopedia-Italiana)/

7 Pistofilo, Bonaventura, *Oplomachia*. Siena, 1621.

8 Brunelli, Giampiero, "GHISLIERI, Federico Fabio", *Dizionario Biografico degli Italiani*, 2000, Vol. 54.

9 https://renaissanceswordclub.com/Desbordes%201610.pdf citing Lepage, André des Bordes ; *épisode de l'histoire des sorciers en Lorraine*. pp. 31-34.

10 Archivio di Stato di Firenze, Mediceo del Principato, Vol. 2949, unnumbered.

1629, Dec. 26th – The German student nation of the University of Siena writes a recommendation for Vincenzo Vannarelli, fencing master, to get a job at Odoardo Farnese's court. Odoardo I Farnese had succeeded his father, Ranuccio I, in 1622. The same German nation at the University had previously employed Capoferro. They say that Vannarelli had three decades of experience in Siena, Germany, and France.[1]

1628-30 – Girard Thibault d'Anvers, a fencing master from Flanders, has a post-humous fencing book published after his death c.1624-27. Thibault had intially studied fencing in Antwerp, but then the family trade as wool merchants led him to Spain. He returned to Amsterdam c. 1610, where he undertook a number of fencing demonstrations, including to Cleves. Eventually he studied mathematics at the University of Leiden, where he composed his book.

Thibault's work shows Spanish-influenced rapier fencing next to fencing with long swords, more Italian styles of rapier, and even musketry. It was lavishly illustrated, and one of the artists was the son of a fencing master, Pieter Bailly, who had stopped tieching fencing at the same University of Leiden in 1602.[2]

1640, Feb. 28th – Francesco Ferdnando Alfieri publishes a fencing book in Padua, where he's the fencing master at the Academy. It includes references to Fabris.[3]

1640, c. Apr. 3rd – Jacopo Monesi writes a fencing book. Published in Florence, he claims to be Master at Arms for the pages at the Tuscan court. The text is dedicated to Francesco Piccolomini, the grandson of Silvio Piccolomini. He claims in the introduction to have been taught by Silvio Piccolomini and Cosimo Paradisi.4

1670, May 23rd – Ferdinando II de' Medici dies. He's succeeded by his son, Cosimo III de' Medici, who rules until 1723.

---

1 Archivio di Stato di Firenze, *Mediceo del Principato*, Vol. 1942, f. 732r.

2 Thibault d'Anvers, Gérard, *Academie de l'Espée*. Leiden, Netherlands, 1630. e.g: https://digitalcollections.universiteitleiden.nl/view/item/1992193

See also Galas, Matt; Eli Steenput. "The Cort Bewijs of Pieter Bailly: A Dutch Rapier Manual, Circa 1602 – 1608." *Arts de Combat. Théorie et pratique en Europe XIVe-XXe siècle*, 2011, pp. 33-76. Ed. by Fabrice Cognot. Paris: Association pour l'Edition et la Diffusion des Études Historiques. & Marika Keblusek. "The Life of David Bailly". *David Bailly - Time, death and vanity.* Waanders Uitgevers, 2023.

3 Alfieri, Francesco Fernando, *La Scherma*. Padua, 1640, e.g: http://data.onb.ac.at/rep/1081E904 Fabris referenced on p. 7

4 Monesi, Jacopo, *Opposizioni et Avvertimenti sopra la Scherma*. Florence, 1640.

1672, Apr. 25th – A work by Pacheco de Narváez, a Spaniard whose effusiveness was only exceeded by his obnoxiousness, is published posthumously. The manuscript was written in 1625. Pacheco identifies many previous authors on the topic of fencing, and shows familiarly with the work of Ghislieri. Amongst others, he names:[1]

Giovanni dall'Agocchie	*Federico Ghislieri*
Camillo Agrippa	*Nicoletto Giganti*
Ridolfo Capoferro	*Achille Marozzo*
Giacomo di Grassi	*Joachim Meyer*
Salvator Fabris	*Girard Thibault*
Alfonso Fallopia	*Angelo Viggiani*

1676 – Thomas Hobbes, English philosopher, composes a work on the physics of fencing for a work being composed by his patron, William Cavendish the Marquis of Newcastle.[2]

1679–81 – Gerolamo Giovanni Battista Christiani, a 26 year old son of a fencing master from Pavia, writes a manuscript on fencing, dedicated to Marchese Giuseppe Maria Visconti. In it he quotes Sir Silvio Piccolomini:[3]

> *"Then, if he has a good eye for the act of extending out the arm to the edge before his sword remains won. But the beauty and the teaching of that most illustrious and excellent Sir Silvio Piccolomini, never sufficiently praised, who said that in no way should he allow his sword to be touched, let alone gained. My Lord Your Excellency knows very well then that when the enemy comes to touch the Sword, so I have taught you to make that beautiful fall underneath with yours, with straightness and straightforwardness, and no other way."*[4]

---

1 Pacheco de Narváez, Luis, *Nueva ciencia, y filosofia de la destreza de las armas, su teorica, y practica.* Madrid, 1672, e.g:
https://play.google.com/store/books/details?id=HXskk57F0pAC

2 Raylor, T. (2000). Thomas Hobbes and 'The Mathematical Demonstration of the Sword.' *The Seventeenth Century,* 15(2), pp. 175-198. doi: 10.1080/0268117X.2000.10555474

3 Cod. Triv. 256 *Compendio di Scherma. 1679,* 5th Dec.-1681, 22nd Jan. ff. 21v-22r.

4 Transcription P. Terminiello:
All'hora habbia ben l'occhio in quel atto che spinge allongara' il suo brazzo per filo addietro la sua Spada rimanera' vincitore, ma il bello come e' l'ammaestramento di quel Ill.mo et Ecc.mo non mai a' bastanza lodato il Sig.r Silvio Picolhuomini il quale diceva che per modo alcuno non si lasciasse, non dico guadagnare, ma' ne ancho toccar la Spada, siche V.S. Ill.ma sa' beniss.mo che quando l'inimico viene per toccare la Spada che li ho' insegnato fare quella bella caschata sotto la sua per rettituda rettitudine, e' non altrimente
https://schoolofthesword.com/Christiani%20Transcription.pdf

1686 – Francesco Antonio Marcelli publishes a fencing book in Rome, dedicated to the Swedish queen. It features a frontispiece of portraits of the Marcelli clan, including a Francesco Marcelli, presumably mentioned by Torquato d'Alessandri in 1609. He mentions many older fencing texts including:[1]

Camillo Agrippa	*Salvator Fabris*
Francesco Alfieri	*Nicoletto Giganti*
Ridolfo Capoferro	*Achille Marozzo*
Giovanni dall'Agocchie	*Joachim Meyer*
Giacomo di Grassi	*Angelo Viggiani*
Marco Docciolini	

1 Marcelli, Francesco Antonio, *Regole della scherma*. Rome, 1686, e.g: https://archive.org/embed/regoledellascher00marc

Bonelli Introduction

The following is an alternate frontispiece and introduction, dedicating the book to Antonino Pio Bonelli, rather than Alessandro Farnese. As far as I'm aware, it was only used for the British Library copy.

RULES OF MANY
KNIGHTLY EXERCISES

Collected by Captain Federico Ghislieri,
*for the instruction of the Most Ill.*ustrious *Sir*
Antonino Pio Bonelli.

⁂

In Parma. Published by Erasmus Viotti. 1587.

With The Permission of the Authorities.

TO THE MOST ILLUSTRIOUS SIR ANTONINO PIO BONELLI,

Son of the Most Excellent Sir Girolamo Bonelli,[1] Marquess of Cassano.

For two reasons, my most illustrious Sir, I have chosen to dedicate this book which I have recently written on the subject of martial exercises to Your Illustrious Self. The first reason is that through this new work of mine, I wish you well, having learned with infinite pleasure of your marriage to the Most Illustrious Lady Ottavia Bagliona, through which your illustrious household may hope for a most fortunate succession. Just as I desire that this labour of mine, under the protection of your name, may live on securely for a long time. The second reason is that by reading this book, Your Illustrious Self, who perhaps has thus far been instructed in the handling of weapons by a good Master, may recognize in it your own virtues, with which you have wished to adorn yourself even in your tender age. Therefore, let not my labour be in vain to you. Although it may be weak in itself, when Your Illustrious Self reads it, you will undoubtedly gain such inspiration that you will hope to live much more than your own strength and merit would allow. With that, recommending myself to your good grace, I kiss your hands.

From Parma, the 22nd of April, 1587.

Your Most Illustrious Self's

Most affectionate servant,

Federico Ghislieri.

Rules Of Many Knightly Exercises.

Collected by Captain Federico Ghislieri, for the instruction of the Most Illustrious Sir Antonino Pio Bonello.

I have undertaken to discuss various exercises of arms, which, although different from one another, all befit the skill and nobility of the knight. Among all types of weapons, it is most fitting for the knight to know how to wield the sword, both because it's more familiar to them than any other weapon and because it is an instrument specifically designed to defend their honour, both in sudden assaults and in close combat engagements. *[... continues like the Ranuccio Farnese version]*

1 Probably Jerome Bonelli.
On 20 July 1572, his relative Pope Philip II conferred on him the marquisate of Cassano sull'Adda, near Milan, with eleven subordinate fiefs, for which he took an oath of allegiance on June 4, 1573. In 1582 he killed a certain Giulio d'Adda with whom he had a long feud about Cassano. See: https://www.treccani.it/enciclopedia/girolamo-bonelli_%28Dizionario-Biografico%29

Transcription

The following transcription and translation is based on the 2018 digitalisation of the Bibliothèque nationale de France's copy of Ghislieri's *Regole di molti cavaglierеschi essercitii*

The publisher used type in ways we don't often see today. I've transcribed:

- ſ as s
- ß as ss
- u as v, when obvious.
- v as u, when obvious.
- ʒ as z when obvious.
- ij as ii where obvious.
- Often a vowel's accent ˜ or ´ is indicating an abbreviated 'n' or 'm'. Unless obvious I've left the original.
- Other abbreviations include ꝑ for 'per' and ꝓ for 'pro'.

Where a line break falls in the middle of a word, the printer often uses '–' to show the continuation of the word on the next line, but not always. I've resolved unhyphenated broken words where obvious.

Where a page break falls in the middle of a sentence, the printer often uses a hanging word or part–word to help follow the text. I've removed that, and added the beginning of the word as an abbreviation on the following page.

I haven't included the header with the page numbers.

fp. §1

REGOLE DI MOLTI
CAVAGLIERESCHI
ESSERCITII.

fp. §2 *Raccolte dal Capitano Federico Ghisliero, per servitio del Ser.[enissi]mo Sig[nore].* RANUCCIO *Farnese, Principe di Parma, & Piacenza, etc.*

fp. §3 [Image of Ranuccio Farnese's coat of arms]

fp. §4 IN PARMA. Appresso Erasmo Viotto. 1587.
Con licenza de' Superiori.

p. i §1 [blank]

p. ii §1 [Allegorical header]

p. ii §2

AL SERENISSIMO SIG[NORO].
RANUCCIO FARNESE
PRINCIPE DI PARMA,
ET PIACENZA.
& C[ETERA].

p. ii §3 *LA gratia, Serenissimo Signoro; che V[ostra]. A[ltezza]. s'è degnata di farmi col servirsi di me nell'occasione del giocar d'armi, m'hà ricordato, che l'essermi adoperato solo con la persona nella medesima occasione sia stato poco, & che perciò mio debito fosse di fare una raccolta, quasi in compendio, di tutto quello, cosi circa la Theorica, come circa la Prattica, ch'iò col mostrare in parole, & con l'operare con la persona, mi sono affaticato volontieri, per ubbidire à lei, d'esplicarle.*

p. ii §4 *Et questo hò fatto non per credermi o di saper perfettamente la professione*

p. iii §1 *dell' adoperar l'armi (che la profession mia veramente non è questa; ma è quella della militia) ò di poter con questo mezo ricompensar la medesima gratia: perche nell'uno sò certo di mancare in molte cose, se ben posso tenermi à molta ventura, & rallegrarmi meco stesso d'esser stato fattura dell Illustriss.[imo] Sig[nore] Silvio Piccolomini, alla cortesissima humanita del quale, in ciò sempre mostrata verso me, nerendo quelle maggiori gratie, che da me si debbono: nell altro, come potre io mai ricompensare à V[ostra]. A[ltezza]. il ricevuto, s'a farlo non mi basterebbono mille fatiche, chio ne spendessi di più?*

p. iii §1 *Ma questo hò voluto sare, accioch'ella habbia segno da me della mia affettuosissima inclinatione, ch io hò di continuamente servirla, & d'adoperarmi per lei con l'opera, come sempre farei con la vita.*

p. iii §1 *Aggiungo à questo un'ardentissimo desiderio, ch'anch'io hò insieme con gli altri suoi fidelissimi servitori. Il qual'è in questo primo fiore dell et à sua di volerla aiutare in quello, che parimente da me si può, per sumministrarle, ancho con questa via da me mostrata, occasione, ond ella possa assuefarsi all'essercitio della persona: il quale quanto*

p. iv §1 *[pro-]fitto faccia in universale alla più sicura conservation della vita, lo mostrano coloro, che particolarmente hanno trattato di questo.*

p. iv §1 *Lo mostrano tanti Gimnasii, & tanti altri luochi già da gli antichi Imperatori (come si legge; & à nostri di ancho se ne veggono i vestigi) fabricati & in Roma, & in altre parti del mondo: affine che & essi medesimi, & i popoli loro ne prendessero essercitandosi in diverse maniere quella vitalità, che dall'essercitio nasce non solo al corpo, ma ancho all'animo: dal qual commodo si vien poi con maggior facilità ad acquistare attitudine per adoperarsi nelle cose cosi di pace, come di guerra.*

p. iv §1 *Et se questo guadagno potiamo far con mill'altre occasioni, sì potiamo molto ben farlo con la via di quest'artè della schrima: dalla quale in un tempo medesimo trahemo due frutti: l'uno è quel, ch'io dissi, il vigore, & la robustezza dell animo, & del corpo. L'altro è ò facendosi da scherzo, ò da dovero adoperandosi, il sapersi difendere contra l'insulto di chi che sià: che questo è il principal fine di colui, che in questo studio s'essercita.*

p. iv §1 *Accetti dunque V*[*ostra*]. *A*[*ltezza*]. *per benignità sua*

p. v §1 *questa mia fatica in quel medesimo grado, che si sogliono accettar le più care cose; che cosa più cara certo à noi non debbe essere, chè quella, dalla quale si possono conseguire, & la conservatione della vita, & la recreatione della mente.*

p. v §1 *Et tanto più volontieri ella dovrà accettar questo mio dono accompagnato da fidelissimo consiglio (che talhora ancho non si disdice al servitore di ben consigliare il Signor suo) quanto che debbiamo tener per certo, ch'ella sarà per far ciò con l'essempio del Serenissimo Signor Duca ALESSANDRO suo Padre: il quale senza dubbio nella medesima età di V*[*ostra*]. *A*[*ltezza*]. *dispensando il tempo in essercitii di questa honoratissima qualità, è finalmente giunto à quel la grandezza di nome, & di satti, della quale hoggi non pure è testimonio tutta la Fiandra; ma ancho tutta l'Italia, & quasi tutto il mondo stupisce dell' incomparabil suo valore.*

p. v §1 *Al quale per gli principii, che tuttavia in V*[*ostra*]. *A*[*ltezza*]. *manifestamente si scuoprono, non ha alcuno, che infallibilmente non speri di veder lei caminar con le medesime vie di pari. La qual gratia piaccia à N*[*ostro*]. *Signor Dio d'essequire*

p. vi §1 [*abon*]*dantemente nella Serenissima persona sua; & à me di compitamente vederla essequita. Et con humilissima riverenza le bacio le mani.*

p. vi §2 *Del Palazzo di V*[*ostra*]. *A*[*ltezza*]. *in Parma gli 22. d Aprile. 1587.*

p. vi §3 Di V[ostra]. A[ltezza]. Serenissima.

p. vi §4 Humiliss[imo]. & devotiss[imo]. Servitore

p. vi §5 Federico Ghisliero.

p. vii §1 [blank]

p. 1 §1 REGOLE DI MOLTI CAVAGLIERESCHI ESSERCITII.

p. 1 §2 *Raccolte dal Capitano Federico Ghistiero, per servitio del Ser.*[*enissi*]*mo Sig.*[*nore*] RANUCCIO *Farnese, Principe di Parma, & Piacenza, etc.*

p. 1 §3 IO mi sono preso à trattare di diversi essercitii d'armi, i quali seben sono tra loro differenti, nondimeno tutti convengono alla sofficienza, & nobiltà del Cavagliero: al quale più d'ogn'altra sorte d'arme stà bene il sapere adoperare la spada, così per esser arma à lui più famigliare, come ancho per esser instruménto truovato propriamente per difender l'honor suo, tanto ne gl'improvisi assalti, quanto nelli steccati da corpo à corpo: & finalmente per esser quella, che ne fatti d'armi hà per fine la vittoria, & ne riporta honore; noi dunque da essa, come da principio di tutti i nostri essercitii incominciaremo: & per non incorrere in confusione, come hanno fatto molti, i quali hanno trattato di quest'arte, avertiremo di

p. 2 §1 [consti]tuire una facultà, la quale si conformi alla natura dell'huomo, quando si truova alterato dalle potẽ[n]ze dell'anima, & sue passioni, che impediscono, che tal'hora l'huomo non possa operare con ragione.

p. 2 §2 IL corpo dunque dell'huomo è có[m]posto di quattro corpi semplici elementali, non che in esso si truovino congiunti i quattro eleméti nelle proprie forme, & nature loro; ma vi sono in quanto vi concorrono con le proprie loro qualità, nelle quali sono le virtú delle dette nature: però che la natura della terra per esser fredda, & secca, genera nell'huomo un'humor detto melancolia, ch'è pur fredda, & secca; l'acqua di natura fredda, & humida fà la flemma; l'aere di natura caldo, & humido fà il sangue; & finalmente il fuoco di natura caldo, & secco fà la colera.

p. 2 §3 Et se bene l'huomo, secondo la natura commune della sua specie, hà il corpo di complessione in maniera temperata, che non declina ad alcuno estremo; contra à quello, che fà la specie de gli animali brutti, i quali havendo le complessioni grandemente inclinate à gli estremi, sono tutti secondo la natura commune à ciascuna delle loro specie naturalmente molto soggetti; come si comprende da certe

p. 3 §1 [lo]ro particolare passioni. Onde veggiamo che tutti i lepri sono timidi, tutti i leoni audaci, tutti i cani iracondi; ma gli huomini soli veggiamo secondo la natura commune à tutta la specie, esser per lo più ne timidi, ne audaci, ne iracondi, ne molto sottoposti alle passioni.

p. 3 §2 Nientedimeno quelle inclinationi, ch'essi non hanno secō[n]do la natura cō[m]mune à tutta la specie, l'hā[n]no secō[n]do la natura particulare di ciascuno huomo, percioche si truovano molti, i quali, perche in essi predomina molto humor colerico, sono naturalmente iracondi, altri, perche abbondano disangue, sono allegri, & audaci per natura; & altri, perche in essi l'humor melancolico supera, sono dolenti, & timidi: & quasi niuno è nel quale siano cosi misurati gli humori, che ne risulti la cō[m]plessione in tutte le parti tē[m]perata, & uguale. Onde avviene, che tutti siamo chi piú ad una, chi più ad un'altra passione inclinati, cō[n] forme alla complessione, che in noi signoreggia.

p. 3 §3 Di quì nasce, che se l'huomo sarà melancolico, si vedranno in lui gli atti, come nati d'elemento terreo, pendenti, ristretti, ansii, & noiosi; si come si vede esser la terra, pendente, grave, & ristretta.

p. 3 §4 I moti dell'acqua, perche sono anchor essi cadenti, se bene non tanto, quanto i terrei, sono nondimeno manco ristretti, & fanno la flemma, la quale corrisponde all'elemento dell'acqua: & ne corpi, ne[i]'

p. 4 §1 quali ella prevale, i moti riescono timidi, semplici, & humili: onde le membra del corpo restano quasi abbandonate, & declinanti al basso; & per la pallidezza, ch'essa infonde ne'volti, fà che alla flemma corrisponde la paura, ò il timore; che vogliam dire; & per la bianchezza cerulea dimostra ne gli huomini il dolore.

p. 4 §2 L'aere hà i suoi moti tendenti all'alto, ma questo non fuor di modo, per esser temperati, & non dilatati, ne affatto storti; come sono quelli del fuoco; & per esser[e]'esso elemento piacevole sono conformi à que sto i moti del sangue dell'huomo; cioè; temperati, modesti, & reali; à i medesimi moti corrispondono perfettamente le passioni dell'animo; cioè; l'amore, dal quale nasce il diletto, il piacere, il desiderio, & la speranza.

p. 4 §3 Il fuoco ultimamente hà di natura in suoi moti tendenti, come si comprende dalle sue fiamme, alla estrema altezza, & in elevarsi tutti si vanno torcendo. Simili a questi sono i moti della colera, percioche sono violenti, impetuosi, & feroci: & essendo à questi moti molto conformi le due passioni, Odio, & Ira, esse perfettamente appariranno in quei corpi dell'huomo, ne quali predominerà questo elemento.

p. 4 §4 Et perche con l'arte si può superar la natura, cercaremo con quella della scherma render il timido audace; & l'huomo audace manterremo tale; &

p. 5 §1 [al]l'iracondo sminuiremo la colera, accioche non sia precipitoso.

p. 5 §2 Et se bene il timore corrompe in gran parte per sua natura il giudicio, & impedisce il consiglio; & se bene l'istesso effetto fà l'iracondia, nondimeno sarà facile al timido far l'audace: la qual cosa si farà, se sminuiremo la paura; laquale, quando non è molto grande, adduce seco diligenza: perche se'l timido con la ragione delle cose si farà capace della facultà dell'armi, esso sarâ più animoso: poiche il mancar la virtù in colui, che teme, è cagione del timore.

p. 5 §3 Parimente all'iracondo moderaremo la colera, & lo faremo più atto ad usar della ragione, se con buoni avertimenti operaremo, che egli non si lasci transportar dalle passioni; ma gli mostraremo, come debba ben bilanciare tutto quello, che può tornarli in aiuto, ò che può dargli impedimento; & che debba porsi ne'pericoli col piè dubbioso, & intrarui pian piano.

p. 5 §4 Sappisi anchora, che per l'imaginatione di quelle cose, che ci fanno parer l'imprese facili, nasce in noi molta speranza, dalla quale poi siamo fatti arditi per quegli avantaggi, che sono assolutamente in poter nostro; come dir le forze, & l'arte della scherma; & di qui animosamente operiamo, & cõ[n] ragio ne otteniamo il nostro fine, ch'è la sola vittoria.

p. 6 §1 *Per mostrar dunque questa facilità nella presente nostra arte, incominciaremo a trattare de[i]'suoi Theoremi: Dopò i quali trattaremo della Prattica, necessariissima a questa professsione.*

p. 6 §2 PRIMIERAMENTE doviamo sapere, che sette circonstanze, over cõ[n]ditioni particolari sono quelle, che si ricercano intorno all'operationi humane: & benche non siano parte intrinseche dell'humane operationi, tuttavia sono sempre necessariamente intorno all'operationi dell'huomo, in modo ch'egli alcuna non ne può fare, che quelle non li siano d'intorno: & la prima è l'operante; la seconda l'opera, over attione operata; sa terza è la materia, intorno la quale s'opera; la quarta è l'instrumẽ[n]to, col quale operiamo; la quinta è in che luoco; la sesta è il modo, secondo il quale operiamo; la settima sarà il fine per il quale si opera.

p. 6 §3 Sarà dunque necessario ad intelligentia di questa nostra scientia dell'armi, che consideriamo le dette circonstanze, delle quali la prima è l'huomo; la secõ[n]da sarà quest'attione di fare alle coltellate; la terza il moto; la quarta la spada per offendere, & il pugnale per difendere; la quinta nelle strade publiche; la sesta il modo, secondo, che offendiamo altri,

p. 7 §1 & difendiamo noi; la settima il fine della vittoria.

p. 7 §2 Truovasi il corpo dell'huomo composto di cosi misurata proportione; & qualunque parte cosi ben rispondente co'l tutto, che gli antichi Architetti dalla proportione medesima cavarono la compositio ne di quasi tutte le cose come di edificar case, chiese, castella, navi, & in somma ogni sorte di fabrica: si come scrivono tutti gli antichi, & moderni, che di ciò hanno trattato; tra quali in particolare è Vitruvio nel principio del libro III.

p. 7 §3 Questa proportione dell'huomo fù giudicata ottima à far questo, ch'io dico, da questi valent[i]'huomini; anchorche nella statura d'esso huomo nõ[n] sia certa, & determinata proportione: perche altri huomini sono maggiori di statura, altri minori: che dovendo esser l'huomo di conveniente statura, havrebbe da esser di sei piedi humani: però che dice il medesimo Vitruvio, che il piede è la sesta parte della statura dell'huomo.

p. 7 §4 A questo proposito Vegetio nel primo libro dell'arte della guerra dice, che il consolo Mario eleggeva i tironi; cioè; i soldati nuovi di sei piedi d'altezza, ò almeno di cinque, & diece oncie, che sono se dieci parti delle dodeci. Onde segue che l'huomo alto manco di sei piedi era tenuto di statura mediocre; & colui, ch'è di più di questa esser altissimo. Onde l'huomo, quando passa il numero di

p. 8 §1 [set]te piedi, è riputato sopranaturale, e chiamato Gigã[n]te, secondo la regola di Marco Varrone, come riferisce Aulogellio nel III. libro delle notti Attiche.

p. 8 §2 Con questo si conforma quello, che Suetonio dice nel la vita d'Ottaviano, parlando della statura di quello Imperatore, la quale era mediocre essendo di cinque piedi, & d'un dodrante, il quale è nove parti di dodeci: ma questa mediocrità non si conosceva, se non quando egli era vicino à qualche persona, che fosse grande.

p. 8 §3 Hanno poi speculando truovato i Filosofi antichi, che la figura circolare, che è la più perfetta di tutte l'altre, si truova nell'huomo perfettamente; la qual cosa esser vera si comprende in questo modo; facciasi che l'huomo si distenda ben con la faccia in sù, & distenda parimente le braccia, & le mani; le gambe, & i piedi quanto più può distendere, & prendasi un compasso, ò la misura d'un piede del medesimo huomo, & si misuri trahendo la misura dall'ombelico, come da centro: che all'hora si vedrà risultare un circolo tondo, & perfetto. Questo hà avertito il medesimo Vitruvio nel predetto libro terzo, & lo di mostra questa figura, che segue.

p. 9 §1 [Illustration of a Vitruvian man in a circle]

p. 10 §1 PLINIO medesimamente nel settimo libro dell'historia naturale à capitoli 17. scrive, che l'huomo è formato di figura quadrata, & angulata; & cio in questo modo, che aprendo esso le braccia, & distendendo le dita si troverà questa maniera formata di modo, che se ne vederà la misura della propria statura d'esso. Onde nasce, che tenendo al modo detto l'huomo i piedi congiunti, & le braccia aperte, vien compreso esser di forma quadrata, di quattro linee uguali: perche una linea li passa per la cima della testa, l'altra per le piante de' piedi, la terza per l'una delle mani, & l'ultima per l'altra mano, come si palesa in questa figura.

p. 11 §1 [Illustration of a Vitruvian man in a square]

p. 12 §1 QUEST[O]' operante hà tre proportioni nel suo corpo, la prospettiva, lo scurzo, & il profilo: la prospettiva è sempre, che stando egli in bilico, mostra tutta la piana superficie del corpo.

p. 12 §2 Lo scurzo mentre che si ritruova nel mezo, fra la prospettiva, & il profilo, mostrando la meta della piana superficie.

p. 12 §3 Et il profilo, quando col dritto piede avanti mostr[a]. la superficie del fianco: come si vede quì dalla dimostratione.

[Illustration of three different views of a man - in prospettiva, scurzo and profile]

p. 14 §1 MA non potendosi passare da un estremo all'altro, che non si passi per il mezo; la prospettiva, & il profilo essendo estremi dello scurzo, non doverà l'operante portar il corpo dall'uno nell'altro, che prima non lo porti nello scurzo.

p. 14 §2 Le dette proportioni del corpo si effettuano per gli moti dell'huomo; i quali si dividono in tre specie; cioè; nel moto di tutto il corpo insieme; nel moto delle braccia; & nel moto delle gambe; i quali essendo i più principali, & che più si confanno al nostro proposito, di quelli diremo, lasciando il rimanente, come quello, che spetta al pittore.

p. 14 §3 Nascono dunque i moti del corpo dalle lungezze, latitudini, & proportioni de membri; & da loro oprimersi, & girarsi, & convenirsi insieme con ragione, & possibiltà; & anchora da loro torcersi, volgersi, & slungarsi, sin'à tànto, che gliè possibile, secondo anchora le incatenature, & chiavi loro.

p. 14 §4 Et per dar qualche regola di questi moti; dico, che essi da otto modi, che tiene il corpo di muoversi, nascono: & sono all'insù, all'ingiù, à destra, à sinistra, stendersi per di là, venir per di quà, volgersi girando, & fermarsi.

p. 15 §1 In ogni moto, ò avanti, ò adietro, è bisogno, che il corpo sia sostentato dalle basi: che altrimenti, per esser la natura de' corpi ponderosi di tendere al centro, caderebbe.

p. 15 §2 Quindi s'è cavato, che i moti del capo sono tali, che à fatica giamai l'huomo non si volta in alcuna par te, che sempre nõ[n] habbia alcuna parte dell'avanzo del corpo posta sotto di se; dalla quale sia sostentato cosi grave peso; overamente, che non ponga dall'altra parte opposta, come una bilancia, alcun membro, che risponda al peso.

p. 15 §3 Percioche il medesimo si vede, quã[n]do alcuno distesa la mano sostiene qualche peso: p[er]che fermato il piede, come fondamento della bilancia, tutta l'altra parte del corpo si contrapone ad aguagliar il peso.

p. 15 §4 Mentre, che in stato si truova l'huomo co piedi pari; & che fà arco avanti di tutto il corpo, & delle gambe; l'estremità del mento resterà à perpendicolo della punta de[i]' piedi.

p. 15 §5 Opprimendosi, ò à mano diritta, ò à mano manca la fontanella, sempre sarà à perpendicolo delli detti piedi.

p. 15 §6 Ogni volta, che il corpo pende da quella parte del piede, che posa, la spalla sarà à perpendicolo del collo del piede; & l'altra gamba sarà per cotrapeso del corpo.

p. 15 §7 Sempre, che se alzerà in alto un braccio, tutte

p. 16 §1 [l'al]tre parti del corpo da quel lato insino al piede seguono quel moto d'alzarsi: di maniera, che il calcagno anchora di quel piede si leverà la dal piano, per il moto del medesimo braccio.

p. 16 §2 Non si slunga mai un membro da una parte, che gli altri non lo seguano; ne per incontro s'opprime, ò si serra, che gli altri non seguano quasi, come linee verso il centro.

p. 16 §3 Le gambe, over basi operano tre sorti di moto; cioè; retto, circolare, & traversale.

p. 16 §4 Retto, ogni volta, che ritruovã[n]dosi il corpo in prospettiva le basi si approno rettamente: come quando si formà un passo.

p. 16 §5 Circolare, ogni volta, che stando fermo il corpo con un piede, descrive con l'altro una circonferenza.

p. 16 §6 Traversale, sempre che si stà in prospettiva; & si fà moto, à mano diritta, overo a mano manca.

p. 16 §7 Le gambe in profilo non si potranno aprire tanto, quanto importa la lunghezza del corpo.

p. 16 §8 Questa linea, overo quantità descritta dall'apertura del compasso del corpo potrebbe esser divisa in infinito: ma à sufficientia di questa nostr'arte, la divideremo in quattro parti; cioè; in mezo passo, in un passo, in un passo, e mezo, & in due passi.

p. 16 §9 Posandosi una gamba l'altra non potrà andarvi avanti più di quanto importa lo spatio d'un piede.

p. 16 §10 Da queste aperture di compasso segue, che in

p. 17 §1 [cin]que modi il corpo potrà trovarsi in stato, over in quiete.

p. 17 §2 Nel primo, quando è in bilico; cioè; co i piedi pari giunti, & che allhora il pendicolare è nel diametro della sua circonferenza; & che in tale stato è atto di moversi à tutte le parti.

p. 17 §3 Nel secondo, quando l'huomo si ferma, con tutto il peso del corpo sopra à un piede, à guisa di base della colonna, il qual stà perpendicolarmente sottoposto alla fontanella della gola, intendendo il collo del piede; & di questa postura fù l'inventore l'antico Policleto.

p. 17 §4 Nel terzo, quando l'huomo si trova in quiete in un passo.

p. 17 §5 Nel quarto, mentre, che il corpo sia in un passo, & mezo, che allhora, forma triangolo equilatero.

p. 17 §6 Nel quinto, quando stà, il corpo compassato in due passi andanti, che si dice passo sforzato.

p. 17 §7 Da queste quieti forma l'huomo sei circoli, due senza, che egli si metta in moto; & quattro doppo, che sarà descritta una delle quantità dette; & che formarà un circolo.

p. 17 §8 Il primo de i due forma l'huomo, quando stando in bilico si descrive un circolo d'intorno à i suoi piedi, il centro del quale è il pendicolare.

p. 17 §9 Il secondo sarà, quando stando l'huomo in detto circolo si descrive un circolo distante dal suo corpo,

p. 18 §1 quanto importa la sua lunghezza; il centro del qua le sarà pur il pendiculare: come si vede in questa figura.

p. 18 §2 [Image of a man standing upright, with a two circles and the diameter of the larger circle drawn on the ground]

p. 19 §1 IL primo de'quattro forma l'huomo, quando egli si truova nello stato della seconda quiete già detta; & quando con la base manca stabile, sopra là quale risiede il peso del corpo, con quella facendo centro, descrive con l'altra base mobile una circonferenza, della quale si serviamo per mantenerci in guardiam.

p. 19 §2 Il secondo, quando si truova in un passo, che stante in quella quiete, fermando il piede manco con l'altro mobile, descrive un'altra circonferenza: & in questa entriamo di moto traversale per uscir fuori co[n]'punti: parimente forma la terza, mentre, che si truoverà nel quarto stato, nella quale entriamo quando si fa la ferita: & la quarta, quando egli si truoverà nel quinto stato di passo sforzato: come dimostra questa figura: & con essa si mã[n]teniamo il nemico lontano.

[Image of a man standing in 'contropasso', with his weight on his left leg. Four concentric circles radiate from the position of his left foot]

p. 21 §1 *Detto c'habbiamo dell'operante, overo artefice, è bene il seguitar di descriver la materia; poiche l'attione di far alle coltellate è per se manifesta.*

p. 21 §2 POICHE il moto, & la materia, intorno alla quale si opera dall'huomo in quest'attione di far alle coltellate, debbiamo sapere, che il movimento, secondo che Aristotile lo diffinisce nel quinto della Fisica, nel nono testo, è una mutatione, overo transmutatione: le specie della quale alcuni vogliono che siano sei; cioè; Generatione, Corrottione, Augumentatione, Diminuitione, Alteratione, & Mutatione di luogo à luogo: nientedimeno l'istesso Aristotile nel prealegato luogo conclude, che tre siano, & non più; cioè; di quantità, di mutation di qualità, & secondo il luogo: delle quali tre specie l'ultima è quella, la quale à noi fà bisogno di sapere per la nostr'[a]arte: il qual movimento non è altro, che quella transmutatione, che alle volte fà movere un corpo da un luogo all'altro; & li termini del movimento sono due instanti.

p. 21 §3 L'instante nel moto, & l'instante nel tempo è, si come il ponto Giometrico nella magnitudine; cioè;
che non ha parte, ma è indivisibile, &

p. 22 §1 [conseguen]temente non è ne moto, ne tempo; ma è ben principio, & fine d'ogni movimento; & d'ogni tempo terminato: come manifesta Aristotile nel sesto della Fisica al testo vigesimoquarto.

p. 22 §2 Siede il movimento fra due quieti: & la quiete, overo riposo, nõ[n] è altro, che privatione del movimento.

p. 22 §3 Et tre sono i movimenti, due semplici, i quali sono il retto, & il circolare: & il terzo composto di questi due, il qual serve alla linea composta.

p. 22 §4 I due moti semplici ò sono naturali, ò violenti.

p. 22 §5 Il movimento naturale è quello, che fanno i corpi gravi da un luogo superiore, ad un'altro inferiore perpendicularmente, & senza violenza alcuna.

p. 22 §6 Movimento violento è quello, che fanno sforzatamente di giù in sù; di quà, & di là, per causa d'alcuna possanza movente.

p. 22 §7 I moti naturali nel principio loro sono deboli; & quã[n]to più vanno continuando il moto, tanto piú divengono di maggior[e] forza.

p. 22 §8 I moti violenti sono di grand'[e]effetto nel principio loro; dal quale quanto piú s'allontanono, tanto più si scema la lor forza, divengono deboli.

p. 22 §9 Et di quattro sorti sono i movimenti di luogo à luogo: l'uno è spingimento chiamato, per cui scacciando da noi le cose, che moviamo, quelle in altra par te spingiamo: come facciamo nello spingere avanti la spada in stoccata.

p. 23 §1 L'altro tiramento è dimandato, per cui al contrario dello spingimento, la cosa à noi tirando, & movendo, la facciamo à noi accostare: come quando à noi ritiriamo la spada.

p. 23 §2 Il terzo portamento è nominato, per cui non da noi discacciando, ne à noi tirando, ma con noi portando, moviamo: si come avviene, quando con la spada ferma in una proportione col moto del corpo nostro portiamo quella hora quà, hora là.

p. 23 §3 Il quarto rivolgimento, ò rotamento si può chiamar quello, per cui in cerchio, movendo alcuna cosa, & quella parte verso noi accostando, & parte da noi rimovendo, giriamo in modo, che tale movimento è quasi di ritiramento, & quasi di spingimento cõ[n]posto: come si vede nelle ferite circolari.

p. 23 §4 Et cinque sono le circonferenze, che col moto forma l'huomo per ferire, poiche nel suo corpo hà cinque centri : Et il primo è nel piede manco , col quale stando fermo, & inalzando il braccio diritto con tutto il restante del corpo in moto descrive una circonferenza, la quale è la maggiore, che egli possa fare: & di questa si forma il taglio.

p. 23 §5 La seconda circonferenza è formata dall'huomo, quã[n]do compassato fà centro del cinto del corpo: & che co'I resto col braccio in alto descrive una circonferenza, della quale, si servono molti in combattere alla bariera.

p. 24 §1 La terza è fatta dall'huomo, mentre che si truova in uno de'cinque stati detti; & mentre che con tutto il corpo fermo facẽ[n]do centro della spalla, & col braccio solo in moto descrive una circõ[n]ferenza, & di q̃[ue]lla si serviamo à fare i mã[n]diritti in stato cõ[n] le basi.

p. 24 §2 La quarta è descritta dall'huomo, quando con tutto il corpo stabile facendo centro del gomito si mette in moto il restante del braccio, chiamato lacerto; & fà circonferenza; della quale si serviamo per cavar le ferite, quando il nemico và alla parata.

p. 24 §3 La quinta, & ultima si fà dall'huomo, mentre che con tutto il corpo stabile facendo centro nella rascetta della mano, con quella sola descrive la minor circonferenza, che egli possa descrivere; della quale si serviamo in far i groppi di mano.

p. 24 §4 *Restaci, per seguitar'[e] il nostr'[o] ordine, che dell'instrumento offensivo, & difensivo diciamo.*

p. 24 §5 L'INSTRUMENTO offensivo è la spada, & il pugnale è il difensivo: & se bene alle volte la spada fà l'offitio del pugnale parando; & il pugnale fà l'offitio della spada ferendo, avviene questo per accidente.

p. 24 §6 La spada instrumento offensivo è compartita in filo

p. 25 §1 buono, & in filo falso, i quali due finiscono inun põ[n]to: Et se bene è corpo contenuto da una superficie; il qual si potrebbe dividere per tutte tre le dimensioni, nondimeno la debbiamo considerare nell'operation sua, come linea.

p. 25 §2 Et in quattro modi si può tener la spada in mano: Nel primo mettendo il detto pollice sopra la sua costa, accioche con questo aiuto si fenda con essa più rettamente: ma in tal modo le punte non si operano bene; & la mano non fà tutta la sua forza, per esser meza aperta.

p. 25 §3 Nel secondo si può pigliar in piano con la palma della mano verso terra. Et ciò si fa per haversela da sentir più leggiera: atteso che la mano col braccio in tale stato opera quasi come appoggio del corpo, il qual fà la lieva al peso della spada: ma in tal modo non si ferisce di taglio; & si giuoca di cambiamento di linee; che è del tutto imperfetto.

p. 25 §4 Nel terzo modo si prende la spada col pugno serrato per avanzare un dito di lunghezza. Il[ ]che nell'instrumento è vero in effetto: ma ogni volta, che si operi la ferita, l'instrumento col braccio hà da formar linea retta, più che sia possibile; il[ ]che non si può far con questa presa: atteso che cosi sempre si forma angolo ottuso, oltre che i tagli per lo più feriscono di piato.

p. 25 §5 Nel quarto, & ultimo modo (il quale è il perfetto)

p. 26 §1 si piglia la spada col pugno serrato; & col dito indice si traversa la croce del fornimento: & in tal modo hà tutte le perfettioni la presa.

p. 26 §2 Questa spada, per la sua percossa, si può assomigliare al cuneo, sopra il quale quanto più sarà grave il peso, che percuote; tã[n]to più si farà la percossa maggiore. Et oltra acciò quanto più sarà lunga la distanza fra il peso, che percuote, & il cuneo; tanto più sarà maggior percossa.

p. 26 §3 Cosi il peso della spada si potrà aggravare più di quello, che sarà per se stesso; & per la possanza movente.

p. 26 §4 Per se stesso: perche ogni volta, che la spada verrà da lontananza, il peso d'essa si aggraverà tanto più, quanto più sarà maggior il moto: atteso che ciascuna cosa grave, mentre si muove, prende più di gravezza mossa, che stando ferma; & di vantaggio più, quanto più da lontano vien mossa.

p. 26 §5 Per la possanza movente in tal modo: perche potendo l'huomo formar cinque circonferenze (come habbiamo detto) conforme à quelle si aggraverà il peso: perche formato che havrà la prima, la quale e la maggiore, si faranno l'altre; & si sminuirà semprè il peso: poiche le circonferenze saranno minori.

p. 26 §6 L'instrumento difensivo, che io posi esser il pugnale, è corpo resistente; il qual'hà per offitio il

p. 27 §1 [difende]re il corpo dell'huomo: & in tre modi si può tener in mano: Nel primo si tiene in mano di piatto, col dito grosso in mezo la lania: & dividendo il corpo in due parti con un semicircolo di moto naturale si difende la parte diritta; & con un'altro semicircolo si difende la parte manca: & perciò và messo in presenza in proportione; affine ch'egli con questi moti faccia le dette difese: il che ha molte imperfettioni: l'una è che quella mano è sottoposta all'offesa: l'altra, se bene muovendo l'ultimo centro solo si descrive una piccola circõ[n]ferenza, tuttavia la linea del pugnale con quella del braccio forma angolo, per il[ ]quale possono entrar molte linee: oltre, che è sottoposto à molti inconvenienti così di dentro, come difuori; & così disotto, come disopra.

p. 27 §2 Nel secondo tiensi in tal modo detto pugnale in mano accostandolo più alle parte diritte; & solo si procura di batter con esso in fuori: accioche prevenga la ferita, avanti che quella arrivi al corpo. Il che è falso: perche in vece di coprirlo, lo scuopre; & è moto violento: oltra che ogni volta, che il pugno forma un poco d'angolo, il pugnale cosi tenuto non vieterà, che quella spada non entri, & anch'egli è sottoposto all'inganno.

p. 27 §3 Nel terzo, & ultimo (il quale è il perfetto) si tiene in mano il pugnale in modo, che con esso si possa

p. 28 §1 [fen]dere bisognando: & di quì co[n]' moti naturali si difende il corpo coprendolo: il quale così tenuto sempre libero dall'offese, non è sottoposto à gli inganni.

p. 28 §2 Oltre che questa linea difendente dominerà sempre per difuori la spada contraria; & co[n]' moti naturali batterà, aiutando sempre le linee alle loro declinationi: perche pochissima forza aggiunge nuovo moto ad alcun peso, che prima si moveva.

p. 28 §3 Si batterà nondimeno l'ultimo terzo della spada, per esser'ella parte più debole in quest'effetto: se bene, quando ferisce, è la più forte: perche muovendosi di moto violento, la prima parte di qual si voglia cosa inanimata e facile da muoversi nell'altra parte: il[ ] che si deve intendere in questo modo, che essendo ogni cosa continua mossa, facilissimo sia il farla muovere obliquamente da quelle estremità, alle quali è congiunto il motore: perche l'altra estremità si transporta con grandissima celerità.

p. 28 §4 Et si come delle cose, che si getano, ò tirano, il moto indebolisce nel fine; così nel fine della cosa continua il moto diviene più debole; & la resistenza minore: & da quella parte le cose più facilmente si spingono, nella quale è maggior debolezza; & perciò minor resistenza: il[ ]che senza dubbio è nel suo fine.

p. 28 §5 Questo pugnale instrumento difensivo ci debbe

p. 29 §1 [for]mar col braccio una linea retta; & l'istessa hà da formar con la superficie del corpo angolo retto: & stante in questo termine la visuale dell'huomo hà da passar per il forte del pugnale; & ferir sopra il debole della spada nemica.

p. 29 §2 Operandosi in questo modo col pugnale, necessariamente con la nostra spada tuovaremo il diametro del circolo del corpo nemico.

p. 29 §3 Et perche sempre deve il pugnale dominar la spada del contrario, si hà d'avertire, che alzando la spada il nostro nemico, noi alzaremo il pugnale in modo, che sempre passando per esso la nostra visuale, veggiamo dominata la spada per di dentro.

p. 29 §4 Il pugnale alle linee rette batte; alle oblique toglie il camino; & alle angolari batte l'estrinseco. Et in tre modi si batte la linea nemica: Nell'uno di volontà, nel secondo di necessità; nel terzo di prosontione: nel volontario si batte in dentro, cavandosi la ferita per la maggior lunghezza.

p. 29 §5 Al necessitato non si batte; ma solo resistendo alla ferita si fà lo scanso col corpo, cavandosi la ferita con voltar il filo falso verso terra.

p. 29 §6 A quel di prosontione si batte con un semicircolo infuori, quandò si truova la spada contraria parata in linea obliqua in presenza sotto il centro del corpo distesa, dovendosi cavar la maggior lunghezza per ferire: & se l'estremo della linea facesse moto

p. 30 §1 [con]trario all'altro, non si cavarebbe la detta lunghezza.

p. 30 §2 Ne si deve con la linea difendente portar la linea contraria al centro, donde nasce: ma si deve, quando ella si truova in esso, batterla: & quando non visia, si deve darle aiuto: accioche più presto di moto naturale arrivi alla sua declinatione.

p. 30 §3 *Hor che dell'instrumento offensivo, & difensivo habbiamo trattato, seguitaremo à dir del luogo.*

p. 30 §4 IL luogo dunque sarà ogni sorte di sito: & perche per il più si viene à rissa, & à duello nelle strade publiche delle Città; le quali spesso sono disastrose sì per le pioggie, sì per gli fanghi; si per esser mal piane: perciò noi ci doveremo accomodare ad ogni sorte di sito: il che faremo, se terremo unito il corpo; & ben compartito il peso d'esso nelle basi.

p. 31 §1 LA sesta conditione, la quale è il modo, secondo che offendiamo altri, & difendiamo noi stessi, consiste principalmente nel'assicurar il nostro corpo dall'offesa del nemico: & perciò l'offesa nõ[n] si farà mai, se prima noi non saremo sicuri della difesa; ne mai si difenderemo simplicemente, se nell'istesso tempo non offenderemo: perche la vera difesa è l'offendere: il che si farà con risolutione; & se saremo sempre i primi ad essequir la ferita.

p. 31 §2 Da queste offese nascono due spetie di ferite: che sono di taglio, & di punta.

p. 31 §3 Il taglio hà tutte queste proprietà; cioè; l'esser più naturale all'huomo, che la punta: atteso che i moti dell'huomo sono circolari: Ricerca maggior portione del corpo contrario: E moto naturale: Per esser più visibile, induce timore. Può anchor terminar in qual si voglia parte.

p. 31 §4 Nientedimeno hà queste imperfettioni; cioè; di scuoprir il destro lato del corpo, & l'istesso braccio; & non è mortale, à quello resistendo l'ossa & l'armi: oltre che per la figura del corpo della spada, la quale per non esser sferica, & perciò di peso non ugualmente grave, molte volte resistendovi l'aere ferisce di piatto: & la spada è soggetta al rompersi.

p. 32 §1 La punta si tira col corpo coperto; & prima arriva, che che sia veduta: & però e più irreparabile; & è di minor moto: & per la forma della spada, la qual è fatta à cuneo, ferisce più sicuramente, & è più mortale: sarà dunque meglio da usarsi, che il taglio.

p. 32 §2 Il taglio poi si divide in tre nature semplici, & due composte.

p. 32 §3 Di queste tre nature di taglio, il primo è il fendente; il secondo il traversale; il terzo il tondo: Et tutte queste tre nature si dividono in diritti, & roversi: i diritti sono quelli, che vengono dalle parti diritte; & i roversi quelli, che vengono dalle parti del lato manco. Et questi diritti, & roversi dividono il circolo dell'huomo in otto parti eguali: come quì si vede nella figura sequente.

p. 33 §1 [Image of a man in a circle with arms and legs outstretched. The circle is divided into eight segments, with vertical, horizontal, and diagonal lines intersecting at the groin.]

p. 32[1] §1 DELLE due nature composte l'una è il diritto ridoppiato, il quale si parte col filo diritto della spada disotto, & và à ferir'alla punta della spalla diritta del nemico.

p. 32 §2 La seconda è lo stramazzone, il qual si fà col nodo di mano à guisa di molinello: & i riversi così si chiamano, perche sono posti à dirimpetto; cominciando dalle parti sinistre, & finendo nelle diritte: & sono delle medesime nature de'mandritti.

p. 32 §3 Il taglio retto, per il suo descenso di moto naturale al centro del mondo, e più grave de gli altri due: poi che forma la linea della direttione, alla quale quanto più s'avicinano nelli suoi moti i pesi, tanto più sono gravi.

p. 32 §4 Oltra di questo occupando, la linea imaginata non concede luogo, per il quale entri in difesa la spada contraria: come fanno i tagli obliqui, i quali sono tutte quelle linee, che segano l'angolo constituito dalla linea del diritto fendente retto, & del tondo fendente.

p. 32 §5 Però questi tagli trasversali quanto piú s'avicinaranno al perpendicolar descenso verso il centro del mondos & quanto più farã[n]no angolo acuto con la linea della direttione, tanto più si avicineranno alla

1 Here the printers made an error, for they repeat page numbers 32 & 33.

p. 33 §1 [per]fettione & per il contrario quanto più se ne allontaneranno, & si approssimaranno alla linea Orizontale del diritto tondo, tanto meno formaranno angolo acuto; & per conseguenza saranno più obliqui; & perciò di minor peso, & forza.

p. 33 §2 Il fendente tondo dalla descrittione di questi due si può intendere à bastanza: & tutto quello, che si è detto de'mandritti, tutto quello si deve intendere de'roversi, poiche sono delle medesime nature.

p. 33 §3 Tutti questi tagli si possono essequire ferendo di polso con un certo moto continuo; cioè; aiutato dal principio, per sino al fine dalla possanza movente: servendosi più tosto l'huomo della forza del suo corpo, che di quella del moto, & del corpo insieme; il che poi s'aiuta con la ragione del segare; che fà la ferita maggiore: & ciò hà qualche apparenza buona: oltre che l'huomo in tal ferire è coperto; & si sbanda manco; & và più concertato nell'operatione.

p. 33 §4 Ma perche la percossa è forza gagliardissima, come dichiara Aristotile nella decimanona delle questio ni mecaniche, concluderemo, che tutte le ferite, affine che siano perfette, saranno aiutate, & dalla forza del corpo, & dalla percossa.

p. 33 §5 La seconda maniera di ferire si divide in tre specie; cioè, in imbroccata, in stoccata, & in punta riversa.

p. 34 §1 [L']imbroccata è quella, che si fa sopra manò; la qual viene da alto à basso di moto naturale; & l'istessa, quando non finisca la declinatione: ma che termini nel piano della spalla; & che formi così nel ferire, come nel riscuotere, triangolo equilatero, si dice imbroccata aventata.

p. 34 §2 Stoccata è quella, che si fà sotto mano di moto violento, spingendosi la propria spada per il suo diritto: la quale anch'ella differisce dalla stoccata affuselata; la qual è così detta, perche fà l'istesso effetto nel ferire, & nel riscuotere, che fà l'imbroccata aventata.

p. 34 §3 Punta riversa è quella, che dalle parti manche si diparte.

p. 34 §4 Perche in qual si voglia sorte di ferita la spada entra inqual che proportione di linea, ci sarà necessario di dichiarare alcuni principii Geometrici.

p. 34 §5 IL punto dunque s'intende esser quello, che per la sua picciolezza è per ogni verso indivisibile: il quale non è quantità; ma termino di quantità: & non può esser compreso da senso alcuno esteriore; ma solo dalla nostra imaginativa. Et questo che segue è il punto: [a dot drawn on the page]

p. 35 §1 La linea è una quantità, con lunghezza, senza larghezza, & senza grossezza, overo profundezza; & per consequenza solo si può dividere per il lungo, essendo per ogn'altro verso impartibile.

p. 35 §2 Et i due termini, overo estremità della linea sono due punti: come si vede in quest'essempio. [A straight horizontal line is drawn]

p. 35 §3 Tre sono le linee; cioè; retta, obliqua, & mista.

p. 35 §4 Linea retta è quella, che da un punto all'altro è distesa, con quella più brevità, che si possa: & è la seguente. [A longer straight horizontal line is drawn]

p. 35 §5 La linea curva, overo torta è quella, che da un punto all'altro è tortamente tirata: com'è questa: [A curved line is drawn.]

p. 35 §6 Sarà puoi la linea mista, quando di queste due sarà composta. [A straight line turning into a wavy line is drawn.]

p. 35 §7 Hora al proposito nostro, la linea retta è quella, che nasce dal sinistro piede col corpo in profilo, & col braccio disteso; & quando l'estremo della spada è al diritto della spalla manca: & sarà anchora la spada in linea retta, ogni volta che sarà in presentia, & che si troverà in stato di ferire rettamente.

p. 35 §8 In linea obliqua si troverà la spada, quando ella sarà obliqua al corpo contrario; cioè; quando si vedrà il corpo della spada, & non la punta.

p. 35 §9 Le linee parallele, ò equidistanti sono quelle, le quali non più da una parte, che dall'altra s'accostano insieme, se ben noi se le potessimo imaginare stendersi in infinito.

p. 38[1] §1 Di quì vi cavo, che la ferita si farà d'angolo retto nel corpo contrario: che così la spada nemica col nostro corpo formarà linea parallela, & hà da cercar parimente la ferita la maggior parte del corpo contrario.

p. 38 §2 L'angolo retto è causato da due linee, che si tocchino inchinate perpendicolarmente l'una verso l'altra.

p. 38 §3 L'angolo acuto sarà quello, che sarà minor del retto.

p. 38 §4 L'angolo ottuso si dirà esser quello, ch'è maggior del retto.

p. 38 §5 Queste linèe si allungano conforme all'aiuto, che gli vien dato da tutto il corpo, & dalle sue parti: come per essempiò, quanto più il corpo s'inchinerà facendo angolo acuto nei suo centro con la linea del la coscia, & con quella del corpo, & parimente quanto più aprirà il compasso, & quanto più causarà, che il braccio con la spada formi linea retta; tanto più la linea della spada diverrà lunga. Et per che queste divisioni si potrebbono mostiplicare in infinito, à sufficientia dell'arte nostra hora basterà, che la dividiamo con le quattro aperture di compasso già dette: come si vede in questa figura.

p. 39 §1 [Illustration representing the reach of lunging to different levels]

p. 40 §1 SI possono però allungare in quattro modi. Il primo più reale è compassando il corpo in profilo; & distendendo il braccio formar linea retta con la spalla manca: nondimeno cotal linea non hà seco la difesa.

p. 40 §2 Il secondo è quando si allungherebbe più, se dopò l'essersi compassato in profilo, si portasse il peso del corpo nella base diritta; & se la manca elevata fosse posta per contrapeso al corpo: con tutto ciò la ferita s'indebolirebbe, & havrebbe troppo grande il moto, & forzosamente bisognerebbe ricadere sopra l'altra colonna.

p. 40 §3 Il terzo è quello, col quale si può portando il piede manco dietro al diritto trasversalmente allungare un poco più la linea: ma questo è moto violento, discommodo, & pericoloso di farci cadere, & non sicuro della ritirata.

p. 40 §4 Il quarto, & ultimo più perfetto di tutti gli altri, è quã[n]do si fà piegando il cinto col corpo in prospettiva, & col compasso aperto; & si forma linea retta: voltando il filo falso della spada verso terra; che così si allunga più la linea, che non si fà in profilo: & questa hà congiunta seco la difesa.

p. 40 §5 Per dar fine alle sette circonstanze, che concorrono in tutte le operationi humane, resta à saper

1 Here the printers corrected their previous error, skipping two page numbers from 35 to 38.

p. 41 §1 [breve]mente il fine, per il quale l'huomo in questa attione si muove; & questo è la vittoria: la quale essendo per se nota, non occorre, ch'altro se ne dica.

p. 41 §2 *Dovendo ogni nostra cognitione discorsiva nascere da qual che altra supposta notitia, donde si possa cominciare à trattare discorrendo, essendo nel rimanente di questo nostro trattato di bisogno d'alcuni altri principii Geometrici, quelli esporremo.*

p. 41 §3 IL Circolo adunque (per cominciar di qui) ò la figura circolare è una figura piana tenuta da una sola linea; in mezo della quale si potrà prendere un punto, da cui tutte le linee, chè si distendessero al giro, overo circonferenza (la qual circonferenza si chiama quella linea, che con tiene tal figura) tutte infra di loro sarebbono uguali: & quel punto si chiama centro.

p. 41 §4 Il Diametro del circolo si chiama ogni [an image of a circle drawn around a point at its centre] linea retta, che passando per il centro di quello, & toccando d'ambe due le parti la circonferenza, divide il circolo in due parti uguali; ò vogliam dire in due semicircoli. [an image of the diameter of a circle passing through its centre point]

p. 42 §1 Il Semicircolo sarà una figura piana curvilinea contenuta dal diametro del circolo, & dalla metà della circonferenza di quello. [an image of a semi-circle]

p. 42 §2 La corda in un circolo si domanda quella linea retta, che divide il circolo in due parti non uguali; & per conseguenza non passa per il centro: come qui si vede. [An image of a cord, a circle where a straight line divides the circle unevenly]

p. 42 §3 Onde quella parte, che rimane maggiore, si chiama maggior portione: & in essa rimane il centro.

p. 42 §4 Quella parte poi, che resta senza il centro di detto circolo, si dice portion minore.

p. 42 §5 Et doviamo sapere, che non solo i corpi, ma anchora le superficie, le linee, & i punti non hanno l'essistentia loro, se non in qualche materia naturale.

p. 42 §6 E ben possibile, che con la imaginatione si cõprendino per loro stessi, & senza applicatione à materia alcuna sensibile: come dir per essempio: In una spada, nella quale effettualmente hà l'esser suo non solo il corpo, ma la superficie, la linea, & il punto; si può ben con l'imaginatione comprendere il corpo contenuto da una superficie, & in essa superficie imaginar le linee, & i punti per se stessi considerati, & senza haver riguardo à ferro, ò ad altra materia naturale.

p. 42 §7 Et in tal guisa consideraremo noi la linea nella nostra operatione astratta dalla materia della spada,

p. 43 §1 [quan]do quella non sarà in presentia; ma quando ella sarà posta in linea retta, allhora consideraremo la linea applicata alla materia della spada.

p. 43 §2 *Hora passando ad altri principii, vi dico, che si come nell'altre operationi humane è necessario, che siano sette conditioni, cosi in questa particolare attione dell'armi le sette medesime concorrono; cioè; volontà, scientia, misura, tempo, occasione, luogo, & peso.*

p. 43 §3 LA volontà è quella, dalla quale dipende ogni nostra attione: attesoche tutte le operationi delle potenze, & di tutti i membri sono, come instrumenti alla volontà, che è il principale agente. La quale è nell'huomo, come un Rè, c'habbia un principal Consigliero, secondo il parer del quale egli sappia di dover fare ogni cosa: & questo è l'intelletto. Hà anchora certi altri sudditi, che sono, come speculatori; se bene alcuna volta riescono buggiardi: & questi sono tutti i sensi esteriori, & interiori. Hà oltre à questi, due altri sudditi, come suoi luogotenenti; i quali hanno da esser pronti in aspettare i comandamenti del Rè, per ubbidirlo: i quali sono la cõ[n]cupiscibile, &

p. 44 §1 [l'ira]scibile, che sono potenze appetitive: l'offitio delle quali è di cõ[m]mandare il movimento alle membra. Ultimamente hà questo Rè un ministro, al quale s'appartiene di dar'essequutione à tutto quello, che da lui, overo da suoi luogotenenti è imposto: & questo essequutore è la virtù motiva, la quale si serve secondo il suo bisogno del corpo, & delle parti d'esso, come d'instrumento.

p. 44 §2 Hora per applicar questo al nostro proposito, dico, che è necessario, che nel fatto dell'armi habbiamo la volontà pronta, con tutti i suoi offitiali; accioche la virtù motiva operi secondo quello, che converrà fare: & sia intenta alla ferita più tosto, che alla parata.

p. 44 §3 LA scientia è cognitione di qualche cosa per le sue cause: & causa si domanda quella, dalla quale viene quell'effetto, del quale essa è causa; & con la quaie si può convenevolmente assegnare sa ragione, donde egli venga.

p. 44 §4 Et si come in quattro modi, & non più, si può con ragione disciogliere ogni dubitatione di qual si voglia effetto, ò la materia di quello assegnando; ò la forma, ò il produttore che io fà, ò il fine, che muove à farlo: cosi di quattro sorti sole saranno le

p. 45 §1 [cau]se de'loro effetti; cioè; la materiale, la formale, la fattiva, & la finale.

p. 45 §2 La material causa è quel soggetto, che stando sotto la forma da quella non si discioglie mai fino, che quella è salva: si come diremo, che in questi nostri effetti della scherma, sia per essempio il moto.

p. 45 §3 La formale è quella figura, ò forma, ò compimento, che dona il modo, & l'essere intrinsico, & appropriato à quel composto, che tale il fà essere, qual si domanda quello, di cui è forma: come per essempio sono le proportioni delle linee.

p. 45 §4 La fattiva si domanda quella causa, da cui viene il principio di quel movimento, & di quella operatione, che necessaria è alla produttione dell'effetto suo: come per essempio, della proportione della linea retta l'huomo, che la fece, è la causa fattiva di quella.

p. 45 §5 La causa finale è quell[a]'utile, overo quel bene apparente, per cui s[i]'induce, & si muove ogni operante à far l'attioni sue, per non operare indarno: come si può veder nell'essempio della ferita retta: il bene apparente di ottenere la vittoria per mezo di quella ferita è la causa finale.

p. 45 §6 Dalla materia dunque, & dalla forma, come da cause intrinseche, & proprie parti essentiali, dipendono tutte le cose composte; cosi artificiose, come naturali: & parimente da quell'artefice, che le fà, &

p. 46 §1 dal fine, che muove à farle, come da cause estrinseche, & forastiere, dipendono esse nella produttione loro.

p. 46 §2 Et in qual si voglia sorte di causa, ò modo di quella, si può cosi la causa, come l'effetto considerare: & questo alcuna volta in potenza, overo prontezza alla produttione; alcuna vosta in atto; cioè; nella produttione istessa: come per essempio ben diremo, che l'huomo, avã[n]ti che faccia la ferita, sia causa fattiva di quella in potenza; cioè; in prontezza, & attitudine à poterla fare: ma non già diremo, chie l'huomo sia causa attuale di detta ferita, finche attualmente non l'habbia prodotta.

p. 46 §3 LA misura per la quale noi ci certifichiamo della quantità della cosa, è quella quantità di terreno, che è frà i due combattenti: & per sin'hora si è cercato di pigliarsi più, & meno cognitione d'essa con la prattica. Et hanno costumato molti, mentre che si rappresentano, di misurarsi: come fanno gli Spagnuoli: ilche è incerto.

p. 46 §4 Ma noi, volendo constituire una misura certa, & determinata, consideraremo il circolo, secondo il quale forma il corpo stante nel primo circolo:

p. 47 §1 [co]me habbiamo dichiarato disopra al suo luogo.

p. 47 §2 Et se bene gli huomini sono ineguali di statura, nondimeno conforme alla loro grandezza descrivereb[be] bono la seconda circonferenza: & presupposto, che il nemico sia maggiore d'altezza, nondimeno non supera mai tanto, che possa avanzare l'avantaggio, che tiene colui, che si truova nel primo circolo.

p. 47 §3 Concludiamo adunque, che necessariamente al nemico, per arrivare al centro di detto circolo, converrà, che con uno de piedi arrivi sopra la circonferenza: perche cõ[n] la lunghezza della spada, la quale è due braccia; & con tutto quel di più, che occupa la mano, & col braccio istesso forma appunto la lunghezza di tre braccia: che tanto spatio è dalla circonferenza al centro.

p. 47 §4 Donde potrà questo tale, che si truova in stato, aprire il suo compasso; & ferẽ[n]do arrivar il suo nemico distante dalla sua circonferenza, quanto importa l'apertura del suo compasso: come quì si vede in questa figura.

p. 48 §1 [An image of the distance of a lunge, shown with concentric circles on the floor beneath a fencer touching another with their rapier]

p. 49 §1 IL tempo hà cosi stretta domestichezza, & congiuntione col movimento, da cui mai non si scõ[n]pagna (peroche non può essere movimento, che non sia tardo, ò veloce; & conseguentemente fatto in più, ò manco tempo) ch'egli è forza, che sustantialmente, overo accidentalmẽ[n]te sia congiunto col movimento; cioè; overo, che sia una cosa istessa con esso; overo, che sia un'accidente intrinsico.

p. 49 §2 Onde essendo chiaro , che se'l tempo fosse movimento, seguiria, che come la velocità, & tardezza conviene all'uno, cosi convenisse all'altro: perche, se bene convien dire, che questo, ò quel moto sia veloce, ò tardo; tuttavia nondimeno non chiamaremo qual si voglia tempo veloce, overo tardo, non potendosi diffinire cosa alcuna per se medesima: resta adunque che almeno debba essergli accidente.

p. 49 §3 Et perche quegli accidenti con la quantità loro, fanno nota, & determinano la quantità d'alcun soggetto, si possono per questo domandar misura di quello.

p. 49 §4 Il tempo è adunque numero, overo misura del movimento, secondo che con due instanti l'uno prima, & l'altro dopoi d'ambe due le parti il

p. 50 §1 [movimen]to determinano. Et tre sono i tempi, passato, presente, & futuro; tra quali del passato à noi non occorre di parlare.

p. 50 §2 Il tempo presente nell'attioni dell'armi non può essere da noi conosciuto se non per accidente, quando succeda, che il contrario operi secondo il costume de gli altri; de'quali s'hà cognitione: perche, se il tempo è la misura del moto, non potremo noi haver cognitione di tal tempo, se non consideriamo prima la natura del moto.

p. 50 §3 Per il qual moto quando è fatto da uno di sua volontà; & che l'altro hà da ricevere il moto da quello, non potrà questo tale, nell'istesso tempo, che l'altro incomincia il moto, mettersi in moto anch'egli: ma si bene il potrà far doppo: & per conseguente colui, ch'è il primo à mettersi in moto, finirà prima che l'altro, c'hà da ricever' il moto da lui. Adunque di questa sorte di misura di moto, la quale è detto tempo fisso, noi non potremo haver cognitione.

p. 50 §4 Dovendo dunque noi haver cognitione del tempo, dovremo in due modi operare: Nell'uno (poiche il moto nasce dalla quiete) consideraremo la natura della quiete della spada nemica; & la quiete parimente dello stato nemico: perche quelle ci dimostreranno il moto futuro; & per consequente haveremo cognitione della sua misura;

p. 51 §1 cioè; del tempo futuro, detto premeditato.

p. 51 §2 Oltra à questo daremo noi il moto al nemico; per il quale lo necessitaremo à far anchor esso un'altro moto; & per conseguente haveremo cognitione del tempo futuro.

p. 51 §3 Avertiremo intanto, che il moto, che noi faremo per dar il moto al contrario, sia di breve tempo, accioche finisca prima che'l nemico se n'accorga: & che volendo egli ferire, habbia da far maggior moto di quello, che faremo noi, in dar il moto à lui: il[ ]che ci sarà facile, se terremo unito il corpo, che possa subito ubbidire alla volontà.

p. 51 §4 Et affine che distintamente conosciamo il tempo: al moto, che noi facciamo con le nostre basi, applicaremo il tempo, che palesa la misura del canto, overo della musica: Et poiche potiamo trovarsi col piede manco avanti in stato, quando da quella quiete partiremo, & col diritto piede contrapassando giungeremo alla quarta circonferenza, la misura di tal moto sarà di otto battute: che tanto importa una Massima.

p. 51 §5 Col piede diritto poi avanti quattro aperture potiamo fare, come habbiamo detto: & la prima è quando ritrovandoci nel primo circolo col diritto in moto facciamo il passo sforzato; & che giungiamo alla quarta circonferenza: che tal moto è di quattro battute, come è una lunga. Ne di questo moto si

p. 52 §1 [ser]viremo, se non per mantenere il nemico lontano.

p. 52 §2 Quando poi formiamo il passo in forza, detto triangolo equilatero (del quale si serviamo in far la ferita) la misura di questo moto consta di due battute: che tanto vale una breve.

p. 52 §3 La misura del passo, che si forma, quando col piede diritto si tocca la seconda circonferenza, che si fà in moto, è d'una battuta, la quale vale una semibreve.

p. 52 §4 In formar noi un mezo passo, faremo la misura d'un moto, che serà di meza battuta: come è il tempo d'una minima.

p. 52 §5 Quella poca apertura di compasso, che si fâ, quando il corpo si truova nella seconda quiete, è di misura di tempo, quanto vale una semiminima: che due insieme fanno una minima.

p. 52 §6 Parimente per dar qualche regola da conoscere il tempo per ferire dico, ch'egli è, quando il contrario fà qualche moto, ò di spada sola, ò con una delle basi, ò sia la diritta, ò la manca, & quando con la spada si fanno, chiamate finte, & provocamenti: à quai moti tutti, ò sieno spingimenti, ò ritiramenti, ò giramenti, ò portamenti, sempre feriremo in quei tempi.

p. 52 §7 Et in tre modi si ferisce, avanti il tempo, nel tempo, & doppo il tempo.

p. 52 §8 Avanti il tempo si chiama ferir di prosuntione: & è

p. 53 §1 quando il nemico è in stato con la spada parata, ferendosi in tal caso alla parte più vicina.

p. 53 §2 Nel tempo si ferisce il nemico, quando egli è in moto, essequendo la ferita; overo parando: & in tal caso si ferisce la parte, che camina in moto, con la ferita, laqual sarà la parte diritta, overo quella, che camina con la difesa, la qual farà la manca.

p. 53 §3 Doppo il tempo si ferisce, quando il moto è per terminare, ritornã[n]do nella quiete; & si seguita la declinatione della ferita: essendo tempo di ferire, quando la spada esce fuori della linea retta imaginata, passando per la nostra visuale.

p. 53 §4 LODANO tutti i savi, che le cose si devono far à suo tempo, & luogo; & non mai fuor d'esso. Però e da considerare, che si come cõ[n]viene aspettarlo, & eleggerlo per operare, cosi è d'avertire di usar diligenza di non lasciar passar totalmente il punto, nel quale è bene à far la cosa, che ci proponiamo: il[ ]che noi chiamiamo occasione, o congiuntura; la quale, quando si perde, rare volte si può ṙ[i]acquistare.

p. 53 §5 Et questa occasione è parte di tempo, il quale hà in se d'alcuna cosa idonea l'opportunità di fare. Et questa in due modi si considera: nell'uno, quando il

p. 54 §1 [con]trario ci provoca con alcuna ferita; overo ci fà finte, ò chiamate: ò quando ci concede la linea retta; & in tal caso non si deve perder la congiuntura di ferire il nemico.

p. 54 §2 Nel secondo modo d'opportunità è quando noi proprii la procuriamo, ma co'[n]debiti modi, & non co'[n]termini detti: che tali sono errori: come se ne verrà in cognitione da quello, che segue.

p. 54 §3 LUOGO non è altro, che la superficie di dentro, & ultima di quel corpo, che contiene: la qual da ogni parte tocchi, & s'accosti all'estrinseca ultima superficie del corpo, che è contenuto.

p. 54 §4 Onde à nostro proposito, sarà la superficie del corpo contrario quella, sa quale toccherà da ogni parte l'estrinseca, & ultima superficie della spada, che allhora in quello ferisce.

p. 54 §5 Et perche intorno à questa materia in due modi si suol dire, che alcuna cosa si truova in luogo; cioè; in un luogo commune; & in un proprio: perciò il luogo commune della ferita sarà tutta la superficie di tutto il corpo: il proprio sarà parte della superficie, la qual per l'ordinario è quella, che manca di difesa, & che più si sporge infuori.

p. 55 §1 Et poiche il centro del corpo in tutti i moti dell'istesso è parte , che stà più ferma, quello, come luogo proprio, feriremo.

p. 55 §2 POICHE, come habbiamo veduto nella compositione dell'huomo, esso hà il suo corpo composto de quattro elementi, de quali gli due gravi causano in lui il peso, c'hà potenza di tendere al basso; & anchora di resistere al moto contrario; cioè; à chi volesse tirarlo all'insuso: & poiche i membri del corpo servono alla volontà, come instrumenti della virtù motiva, non potremo con gli detti ubbidire, se il peso non sarà compartito in essi, secondo che il bisogno ricercherà.

p. 55 §3 Habbiamo dunque da sapere, che il peso del corpo è compartito sopra le due colonne: & quello si può hora in una sola, hora in un'altra porre, & scaricare. Ma questo s'havrà da far con ragione: perche essendo queste due colonne gli estremi del pendiculare, il qual è nel loro mezo; & non potendosi passar da un'estremo all'altro, che non si passi per il mezo, converrà che il peso con queste due possate si transporti nelle colonne; come, se trovandosi il peso su la base manca, si vorrà transportare

p. 56 §1 nella diritta, allhora bisognarà che prima si posi nel mezo; cioè; nel pendiculare; & poi si scarichi nella diritta: & questo parimente si farà dalla diritta nella manca.

p. 56 §2 Et perche la parte del corpo, la quale haverà da far alcun moto, overo retto, overo obliquo, non potrà ciò fare commodamente, se non sarà alleggerita dalla parte del suo peso: però sempre la colonna, la qual havrà da star ferma, come se fosse la gamba d'un compasso, sosterra tutto il peso, accioche l'altra alleggerita possa ubbidire alla volontà, essequendo il moto, che gli sara commandato.

p. 56 §3 Et caminandosi dispostamente co'passi naturali si terrà il peso unito in atto; accioche si possa riporre sopra una delle due basi conforme al bisogno.

p. 56 §4 Parimente l'huomo, volendo guadagnar terreno à banda diritta, overo manca, dovera moversi portando il peso nel mezo: & poi quello nella base diritta scaricando raccorre sa base manca nel primo circolo: che con quest'ordine procedendo sempre si truoverà in atto di rispondere alla ferita. Però non doverà mai nello stretto far moto seguito con tutte due le gambe; salvo che quando ei camina.

p. 56 §5 Et accioche l'huomo possa ottenere le dette particularita, doviamo sapere, che il secondo circolo, il qual egli forma in moto, si divide in quattro angoli retti con due diametri: & stante l'huomo nel

p. 57 §1 primo suo circolo; cioè; in billico; & convenendogli uscire, ò à man diritta, ò à man manca, ò avanti, ò in dietro, questo farà sempre movendo solo l'una delle basi, & con l'altra stabile restando nel centro del primo circolo.

p. 57 §2 Et è da notare, che mentre ci troviamo co'piedi nel primo circolo, & che usciamo con una base dal primo; & che entriamo nel secondo, che si fà in moto, allhora quel circolo secondo divideremo in quattro figure piane trilatere, chiuse, & contenute da tre linee rette: & considerandosi ciascuno triã[n]golo da se solo, ogni volta che si troverà l'huomo posto in uno d'essi, & che vorrà entrare in uno degli altri, doverà unendosi ritornare nel primo circolo, che si fà in stato: & poi conforme al bisogno anderà in una delle dette quattro figure; ilche tutto si dimostra con la figura, che segue.

⁂

p. 58 §1 [Image of a fencer. Drawn on the ground are two concentric circles centred on his rear left foot. A square is superimposed over the two circles.]

p. 59 §1 *Poiche habbiamo dichiarato le sette circonstanze, le quali concorrono all'operationi dell'armi, doviamo sapere, che à voler, che un'attion nostra dipenda da virtù, è necessario, ch'ella habbia quattro conditioni, cioè, ch'ella sia spontanea, consultata, eletta, & voluta.*

p. 59 §2 SPONTANEAMENTE fatta quell'attione s'intende, che noi di nostra propria volontà facciamo: & per il contrario, quelle operationi, che noi facciamo non di nostra volontà, si possono dire non spontanee.

p. 59 §3 Da questo segue una regola per noi: che l'huomo nell'attioni dell'armi farà tutte le sue operationi volontariamente, essendo il primo à cõ[m]mettere risolutamente: & per consequente operarà in maniera, che il suo contrario habbia da far il tutto contra sua voglia, per forza necessitandolo al pararc.

p. 59 §4 La consultatione; come determinò Aristotile; è intorno à quelle cose, le quali possono cadere sotto il consiglio humano.

p. 59 §5 Onde noi in questa nostr'arte ben bene consultaremo tutti i mezi, che ci possono portare al nostro fine, che è la vittoria.

p. 60 §1 L'elettione non è altro, che un consentimento, overo assenso consultatiuo di quelle cose, che sono riposte in noi: percioche essendo una cosa prima consultabile, & poi eleggibile, se prima sarà consultata, & per buona giudicata; verrà ragionevolmente ad essere eletta.

p. 60 §2 Talche da queste due conditioni, ch'io congiunsi insieme, constituiremo una regola, che doppo l'haver consultato i mezi dell'offesa, & difesa, eleggeremo la miglior proportione di ferita, & di difesa, & quella di continuo operaremo.

p. 60 §3 Il voler nostro è pronto ad haver riguardo, à quel fine, che non solo è vero bene, ma è apparente: per l'acquisto del quale debbiamo poi spontaneamente in noi consultare de mezi, che à quello ci posson condurre; & quelli finalmente eleggendo virtuosamente operare: come faremo noi nel cimento dell'armi, che si fà da corpo à corpo; se havuto riguardo al vero bene, il quale in questa attione è il difendere se medesimo, & offendere il nemico, le cose consultate, & elette animosamente metteremo in essecutione, che altrimente facẽ[n]do s'annulla tutta l'arte dell'armi.

p. 60 §4 Et dato, che da un punto ad un'altro non si dia se non una linea retta: & essendo, che trovandosi una linea nel soggetto, l'altra non vi possa entrare, per non poter due corpi essere in un istesso tempo nel

p. 61 §1 medesimo soggetto; resta che ogni volta, che due verranno al combatter fra loro, necessariamente nascerà il circolo della distanza: il quale partito dal suo diametro, & formando ciascuno di loro un punto, chi prima di loro metterà la sua spada nel diametro, che per conseguenza vien ad esser la linea retta, sforzerà il contrario à passar per essa, in modo che haverà cosi l'offesa per sua, come anchora la difesa: poiche la spada nemica si fà parallela al suo corpo. Et accioche questo meglio s'intenda, si dovranno ben considerare queste figure, che qui sotto seguono.

p. 62 §1 [Image of two men, standing on the edge of a circle and joined by a line indicating the diameter of the circle. The figure on the right has his sword extended, touching the other.]

p. 63 §1 NEL che fondandosi la scientia dell'armi, quindi è avenuto, che varie sono state le operationi di questi, che per sin'hora hanno trattato di tal materia: gli quali overo si sono messi in essa, overo hanno dato il luogo, accioche fe rendo il nemico essi potessero in tal tempo assicurarsi parando; & potessero mettendo la loro spada nella linea retta ferire.

p. 63 §2 Coloro, i quali hanno messo la spada avanti in linea retta, ciò hanno fatto per la ragione detta, ch'è principalissima; & per tenersi coperti dalla spada; & per tenersi il nemico lontano: da che sono nati tanti, & si diversi modi per guadagnarla.

p. 63 §3 Et per il primo, secondo ches'usa di fare in Hispagna, & ancho in Italia, è il guadagnarla compartendo la spada in tre parti: delle quali la prima verso il pugno è la più forte, perche e più vicina alla possanza movente: & l'ultima è debole, poiche è lontana dalla possanza movente: ma avertasi, che io dico debole nel parare, se ben è fortissima nel ferire.

p. 63 §4 Con questa divisione procedendosi, si averta disopra metter la spada sua à quella del nemico nel debole, tenendosi il braccio raccolto, & il pugno della mano posta al piano, overo un poco più basso di quello del nemico. Di modo che aspettando, che il

p. 64 §1 contrario la liberi, nel medesimo tempo si guadagni la linea retta col slongare il braccio. Dovendosi tener per regola l'haversi in un tempo ad andar alla spada; & con un'altro al corpo nemico.

p. 64 §2 In difesa di questo, ch'io hò detto, solevasi nel tempo, che quella spada era in moto, affine di soprastare, cavare in detto moto, & liberar la spada, ò con quella sola facendosi moto, overo col giuoco di scossa, il qual è portar il corpo da una base nell'altra; overo ritirandosi il piedè manco adietro: & cosi procedendosi di mano in mano l'uno contra l'altro si operava. Ma questo più tosto per far da scherzo, che per far da dovero, è buono: attesa la maniera de gli huomini, essendo essi ineguali, & di grandezza, & di forza: come anchora sono le spade ineguali & di lunghezza, & di peso: & oltra à questo l'huomo alterato non discerne queste minutie; & essendo questo giuoco fondato sopra principii incerti, sarà anchor esso incerto.

p. 64 §3 Alcuni ꝑ[er]remediar'à molti di quaisti inconvenienti ritrovã[n]dosi in detta linea posti cõ[n] la sua spada; quã[n]do il nemico và ꝑ[er] guadagnarla, in q̃ [ua]l tẽ[m]po essi portano la sua spada fuori della linea retta; & rincontrando la spada nemica fuori di forza; e poi mettẽ[n]do la spalla mã[n]ca senza distaccar la sua spada da q̃ [ual]la del nemico in quell'angolo cõ[n]stituito dalle due spade cõ[n]trapassano col piede manco, & feriscono di pũ[n]ta riversa.

p. 65 §1 Molti havendo considerato, che mentre che si guadagnava il debole della spada nemica, essi potevano in questo tempo esser feriti, col cavar la spada il nemico hanno tenuto quest'ordine nell'andar alla spada nemica, di portar il corpo verso quella, acciò che se'l nemico liberasse la spada, & che ferisce, nõ[n] trovasse luoco dove ferire: onde essi in quel moto mettevano la sua spada nella ferita in linea retta.

p. 65 §2 Ma quando la spada nemica fosse stata alta, davano i medesimi occassione al nemico, accioche ferisce: & questo facevano con certe tagliate di spada con l'arme accompagnate, & unite, con l'istesso moto del corpo à mano diritta, overo à mano manca; & ferendo il nemico essi in quel tempo è legavano la spada nemica con un molinello, & ferivano col pugnale; overo ferivano mettendo in quel tempo la sua spada in linea retta.

p. 65 §3 Si è costumato parimente, per distorre questa linea retta dalla sua postura, far delle finte ò per di dentro, ò per difuori con punte, overo con taglii, per impaurire quel tale, che si trovasse con la spada in presentia: accioche uscendo con la sua spada dalla linea per parare, concedesse luoco, nel quale potesse mettere il nemico la sua spada: ma questi tali, che fanno queste finte, prinia essi commettono un errore per farne far un'altro al nemico: i quali

p. 66 §1 [er]rori poi ne' fatti dell'armi si pagano con la perdita della vita.

p. 66 §2 Et perche colui, che si truoverà in questa postura, con ogni poco moto si manterrà sempre nel detto vantaggio, se sempre terrà fisso la punta della sua spada nella spalla diritta del nemico; & dovunque ella anderà, la seguiterà con detta postura: di quì s'imaginorono molti di non ritrovare quella spada, premendogli il suo debole, ma d'adombrarla con la loro; & accioche havessero distantia per ferire, ciò facevano col piede manco avanti: & avertivano di mettere il piede mã[n]co al diritto appunto del pie de del nemico, & la spada distesa in linea retta à livello della nemica; & arrivati in distantia, & col piede diritto crescendo ferivano.

p. 66 §3 Ma colui, che si truoverà posto in detta linea, se farà moto solo col piede diritto traversalemente uscendo, ferirà il nemico: perche quando egli sia ricerco per difuori, potrà ferire di punta scavizzata, cosi detta da questi maestri di scherma, overo di gobba: & quando la provocaremo per di dentro, si potrà ferir di punta riversa, overo d'incapocchiata.

p. 66 §4 Altri si sono fondati sopra l'offese, & difese; & in vario modo hanno proceduto; overo stando il nemico in guardia, hanno procurato di far qualche sorte di ferita o di taglio, o di punta per provocar il nemico; accioche scõ[n]certandosi essi ò con la

p. 67 §1 [pa]rata ò con la ferita, possino essi in quel tempo parare, & poi ferire.

p. 67 §2 Overo, quando hanno truovato la spada in presentia, quella hanno sconcertata, ò battendola nel suo debole col diritto filo all'ingiù, ò col falso filo di sotto in sù: accioche essi potessero in quel tempo ferire: nel qual modo procedendosi le parate, & ferite si riducono all'infinito: come si può vedere ne libri, che di cio hanno trattato.

p. 67 §3 Quelli, i quali danno un luoco discoperto, ciò fanno per dare occasione al nemico di ferire: accioche possino parare, & ferire: ma à questi le finte nocio no grandemente: & oltre di questo il medesimo modo si riduce all'infinito.

p. 67 §4 MA noi, per non incorrere in queste sorti d'offese, & di difese, non ci cureremo ne di mettere la nostra spada in linea retta; ne, essen dovi quelsa del nemico, cercaremo di quadagnarla co' detti termini: anzi secondo la debita ragione operaremo. Et perche la vera scientia consiste in conoscere le cause desse cose; le quali conosciute, & rimosse, che sono, si rimuove anchora l'effetto, considerare mo la causa di tal linea; la qual è il punto della spalla nemica, & il punto del

p. 68 §1 [no]stro corpo, ove mira detta linea: & ogni volta, che si rimoverà il punto di detta linea nel nostro corpo estremo, si rimuoverà per consequente l'effetto di tal linea.

p. 68 §2 Et in duè modi si puo rimuovere detto punto: Nel l'uno, non si perdendo terreno; & col moto traversale uscendosi ò à man diritta, ò à man manca: Nell'altro, quando si perde terreno col ritirarsi un passo, & col ferire in quel tempo medesimo.

p. 68 §3 Il portar fuori della linea retta il punto si fà in questo modo; cioè; havendo noi compartito il corpo in due parti eguali dal diametro del nostro circolo, in ciascheduna parte noi constituiamo un punto; & ò di nostra volontà uno d'essi mettiamo per estremo della linea; overo il nemico è quello, che mette la sua spada in uno d'essi, & quando di nostra volontà lo porremo, sempre sarà il diritto, per cavarne la ferita più sacile, & commoda: perche cosi guadagnaremo un passo di terreno.

p. 68 §4 Quando il nemico poi sarà quello, che metterà la sua spada in uno de' nostri due punti, se per aventura la metterà fiel diritto, usciremo col piede manco; se nel manco, usciremo col piede diritto: che cosi operando si rimuove l'essetto di quella linea nemica.

p. 68 §5 Il rimuovere il pũ[n]to, oltra che annulla la linea nemica, à noi genera poi un'altra linea, nella quale

p. 69 §1 [met]tendo la nostra spada potiamo ferire; & cosi ci salua il nostro corpo, il qual fugge le prime ferite; & di più, ci assicura dalle seconde.

p. 69 §2 Quando poi si rimuove il punto con la perdita del terreno; cioè; quando quello allontaniamo dalla spada nemica, avertiremo, che questo non si fà se non ferendo il nemico: il quale quando avenga, chè ci ferisca le nostrem parti diritte, allhora noi con un passo indietro le ritiraremo; & nel tempo istesso feriremo il suo braccio: ma ferendo esso le parti nostre manche, col manco piede perderemo il terreno, & in quel tempo feriremo. Ma di questo modo noi non ci serviamo, potendosi in miglior modo operare.

p. 69 §3 Et poichè ogni sorte di causa, & ogni sorte d'effetto si può considerare alcuna volta in potenza alla produttione; & altra volta in atto: cosi quella linea consideraremo ò in atto, che sarà, quando la spada sarà già messa in essa; overo in potenza, quando essa non vi sarà; ma sarà bene atta à venirci: attesoche tutte le linee in qual si voglia sorte di proportione, ch'elle siano, ferendo capitano alla linea retta: come si vede in queste figure, che seguono.

p. 70 §1 [Illustration of two men, on extends a sword at the other's face. The two stand on a line on the ground. A parallel line connects their torsos. A line extends from the tip of the extended sword to a circle, divided into eight sections, superimposed on the other's chest. This is a call-back to the illustration on p. 33]

p. 71 §1 DISTINGVEREMO, adunque, che il nostro contrario ò sara in stato; cioè; in quiete; ò da quella partendo, si truoverà in moto.

p. 71 §2 Mentre che egli sarà in quiete, ò terrà la spada in linea retta; cioè; nella produttione istessa: overo, havendola fuori della detta linea, la terrà in potentia, ò vogliamo dire prontezza di metterla
nella detta linea.

p. 71 §3 Quando egli terrà la spada nella produttione istessa; cioè; nella linea retta, dovremo considerare, che conforme all'apertura del suo compasso abbassandosi la spalla diritta la spada anderà al livello con la spalla; & per consequente il punto della nostra superficie sarà al diritto di quella linea: la qual linea sempre si chiamerà retta, insino che in detta proportione ella arriverà al diritto del centro del corpo nostro.

p. 71 §4 Mentre che il nostro nemico terrà la spada in potenza per metterla nella linea retta, si chiamerà in linea obliqua; la quale ò sarà avanti; cioè; in presenza, ò doppo; cioè; fuor di presenza.

p. 71 §5 La spada si dirà obliqua in presenza in tre modi: Nel primo, quando ben sara in presenza nel diametro del nostro circolo: ma si truoverà poi sotto il centro del nostro corpo: Nel secondo si dirà obliqua

p. 72 §1 la spada nemica al nostro corpo, quando ambidue saranno nel diametro del circolo della distanza; & che la punta della spada nemica sarà fuori del nostro corpo, ò alta, ò bassa, ò in qual si voglia sorte di proportione: & questo sarà tenendo al contrario il pugno della spada alle sue parti diritte, overo alle manche: Nel terzo la spada contraria è obliqua al corpo nostro, mentre ch'ella forma angolo col braccio; & che la sua punta supera il centro del nostro corpo.

p. 72 §2 Fuor di presentia, overo indietro in due modi il nemico potrà tener la spada: ò col piede diritto avanti, ò col manco; col diritto la terrà ò raccolta, & bassa; ò alta, & in atto d'operare una coltellata: quando poi la terrà col piede manco avanti, ò la terrà alta in guardia di falcone; ò bassa in guardia chiamata da molti coda longa, & larga.

p. 72 §3 Et tutte queste linee si possono mettere in infinite proportioni: & quindi è avenuto, che tante guardie si sono formate: perche da tutti i principii delle ferite; da tutti i suoi mezi; & da tutti i suoi fini si possono formar posture: ma accioche non ci confondiamo, avertiremo solo à queste proportioni, c'habbiamo dette; nelle quali forzosamente il nemico si truoverà: importando poco, ch'elle siano ò un poco più alte, ò un poco più basse.

p. 73 §1 HORA ritornando alla dichiaratione dell'arte nostra, dico, che ritruovandosi il nemico in qual si voglia sorte di quiete in postura, doveremo per operare con scientia considerare le quattro cause, le quali cagionano gli effetti: Et primieramente la causa efficiente, che è l'huomo: & questa è la più generale, & più remota nell'attione di questa postura: ma la più propinqua, & particolare sarà l'atto, nel quale ess'huomo si truova; il qual atto ci dimostra l'effetto, che può nascere da quello.

p. 73 §2 La causa formale, la quale è il moto in generale dalla sua quiete, ci denota la misura del moto particulare, il qual'è per mettere quella in effetto.

p. 73 §3 La causa formale, la quale è la proportione della spada nemica, ci palesa l'effetto della sua ferita.

p. 73 §4 La causa finale voglio, c'hora in quest'effetto sia quella, che noi consideriamo intorno alla linea della spada, la qual ferendo hà per fine il colpire nel nostro corpo. Però la final causa sarà il punto della nostra superficie.

p. 73 §5 Et ritrovandosi il nemico con la spada in linea retta; ma sottoposta al nostro pugnale, come se egli fosse in guardia terza, & quarta, allhora battendo quella linea, & nell'istesso tempo ferendo

p. 74 §1 [mettere]mo la nostra spada nella linea retta: ma con la spada sola usciremo col piede manco ferendo.

p. 74 §2 Mentre poi il nemico terrà la sua spada contra di noi in linea retta, in stato col piede diritto, overo man co avanti, come in guardia di testa, ò d'intrare, ò di faccia; ò col diritto quanti in guardia prima, & seconda, terremo per regola d'uscir fuori da quel diritto di moto traversale.

p. 74 §3 Quando poi la spada nemica sarà obliqua al nostro corpo; cioè; nel suo primo modo sotto il centro del nostro corpo, dovremo col piede manco ferire in quell'istessa battuta, per formar lo scanso del corpo: & quì parlo di spada sola: perche se havremo il pugnale, lo metteremo alla linea imaginata, & fe riremo col nostro solito modo.

p. 74 §4 Se poi il nemico haverà la spada obliqua al nostro corpo nel secondo modo (che in questa proportione vederemo il corpo della spada, & non la sua punta) sempre guadagnaremo alla nostra man diritta, & fuggendo quella lunghezza, & ferendo metteremo la nostra spada nella diagonale.

p. 74 §5 Ma titruovã[n]dosi obliqua la spada nel terzo modo, dovremo distinguere, che se il punto di quella linea angolare sarà al diritto del punto nostro diritto, allhora contra quella operaremo, come se fosse in linea retta: ma se fosse co la punta della spada dirimpetto al nostro punto manco, la segaremo,

p. 75 §1 [cac]ciando la nostra spada in linea retta nella ferita.

p. 75 §2 Caso poi, che'l nostro contrario tenesse la spada adietro, overo fuori di presenza col piede diritto avanti, come fanno alcuni, quando tengono la spada raccolta per far l'inquartata: overo quando sono in guardia alta in coda lunga, & larga; overo col piede manco avanti in guardia di falcone; overo in coda lunga, & alta: & sia pur finalmente in qual si voglia sorte di proportione, pur che sia fuori di presentia; curaremo sempre di combattere col corpo; & quello cercaremo di moto con le nostre parti diritte, essendo esso (come dice Aristotile) natule à tutti gli animali.

p. 75 §3 *Hora, c'habbiamo dichiarato le proportioni delle linee, quando sono in stato, diremo di quelle, che sono in moto.*

p. 75 §4 DOVIAMO dunque considerare, ch'elle partono da una quiete, & si vengono à riposare in un'altra: perche tutte le ferite partono dal centro della spalla, ch'è origene della linea retta; & nella ferita ritornano pure in detto centro.

p. 75 §5 Et perche tutti i moti dell'huomo nelle ferite sono

p. 76 §1 [mi]sti; atteso che per volere (come per essempio) far una stoccata, conviene prima, che egli faccia moto circolare, per causare il retto: & parimente per che l'huomo formando una ferita ciò fà di moto violento, & naturale (ma il violento è di maggior vehementia del naturale) però distingueremo, che'l moto hà principio, mezo, & fine.

p. 76 §2 Et accioche questo trattato sia meglio inteso, io, in questa parte, che segue, pospongo il moto obliquo, & violento, che fà l'huomo, quando prepara il braccio, & lo porta in alto, & lo riposa nello stato, per formar ò taglio, ò punta, se bene è tempo di ferire il nemico: solo nondimeno intendo di dire del moto, che si fà, quando il braccio partendo da quella quiete declina in moto alla ferita; & questo moto è quello, che è (come habbiamo detto) composto di violento, & naturale; violento rispetto alla forza della possanza movente, la quale spinge, & violenta quel peso al suo descenso: naturale poi rispetto al peso del braccio, & della spada, che di sua natura tende al centro della terra.

p. 76 §3 Onde il detto moto hà il suo principio debole; il quale è quando incomincia à discendere: hà poi il mezo forte, che è quando il braccio giunge al livello della spalla: nel qual piano è la maggior percossa, che possa far l'huomo; perche quiui termina la

p. 77 §1 [violen]tia della forza dell'huomo: hà finalmente il fine fuor di forza, ò meno forte; il quale è quando il braccio eccede quel piano della spalla: perche il moto diviene più naturale, poiche la violentia vàcessando, essendo il braccio poco aiutato dal resto della possanza movente.

p. 77 §2 Et distinguendosi questi tre termini differentemente, doveremo contra ciascheduno di loro operare: & perche il parare è infimo (di quello della spada parlando) non si dovrebbe fare: perche non si concedendo il vacuo nella natura, se l'uno sarà il primo à ferire, & che l'altro pari, continuando colui, che è primo à ferire, il moto, sempre terrà sottoposto l'altro alla parata: onde in due modi si schermiremo dall'offese nemiche.

p. 77 §3 Nel primo, se si trovaremo in distantia, potremo nel principio del moto ferire: che cosi intraterremo il contrario: il quale percosso (per la natura, che repugna al male) non seguiterà il moto incominciato. Et questo si dirà ferire avanti il tempo, mentre che il moto è debole.

p. 77 §4 Et avertiremo molto bene, che questo modo di difendersi si faccia risolutamente, mettendo bene il corpo in profilo, con distendersi; & ferendo alle parti diritte, & nemiche: & questo fin tanto, che'lpeso del braccio, overo quello della gã[m]ba mobile in quel moto starà à pióbo verso il centro della terra.

p. 78 §1 Di piú: le spade, le quali tagliano, & pungono, fanno differente effetto di quelle di marra: perche, se noi ferissimo il nostro contrario, quando il suo peso è già in declinatione con la spada, che punge, & passa, bene offenderessimo, ma il nemico potrebbe anchor'esso ferir noi.

p. 78 §2 Et la ragione è questa, che se'l peso del corpo, il qual si trovava à piombo col suo diametro nella piana superficie terrea, fosse rimosso dal pendiculare, & dal punto; cioè; dalla quiete: & che'l peso fosse in declinatione al suo natural cẽ[n]tro, non si potrebbe ritirare quel peso all'insù: poiche tutte le cose difficilmente si muovono verso la parte contraria alla sua intrinseca inclinatione.

p. 78 §3 Et in confirmatione di questo, l'anima dell'huomo, la quale in esso è una sola, non può in uno istesso tempo essere grandissimamente intenta à più cose: come dire con la ragione à commandare i movimenti alle mani, & all'altre membra; & con le sensitive virtù ad apprendere il dolore delle ferite.

p. 78 §4 Però concludiamo, ch'egliè ottimo il ferire il nemico avanti, ch'egli habbia il peso del suo corpo nella declinatione.

p. 78 §5 Nel secondo modo di schermirci dalle linee in moto, in due modi potremo cio fare: & nel primo si potrà perdere il terreno allontanando il punto del nostro corpo; all'uno de' due modi, c'habbiamo

p. 79 §1 detto, quando si perde un passo di terreno, per uscir dal diritto con i punti: & si ferirà al braccio contrario: overo si contrapassarà col diritto, uscendo col punto di presentia: & si ferirà pur al braccio di stramazzone. Et questo si dice ferire in tempo nel moto forte.

p. 79 §2 Nell'altro si lasciano passar le ferite à vuoto; & subito che la spada passa per gli raggi della nostra visuale con quelle sorti di ferite, che à noi tornano più bene, doppo il tempo nel moto debole si ferisce.

p. 79 §3 *Ma perche il più de gli huomini non hà questa cosi sottil cognitione di offendere altri, et di difender se medesimo: perche gli huomini mossi dalla natura, parano le ferite nel mezo del suo moto; cioe; nel forte, quando il braccio giunge al piano della spalla. Però noi discorrendo delle parate eleggeremo la migliore.*

p. 79 §4 ET in quattro modi si para: Nel primo si aiutano tutti i moti alla sua declinatione: perche tutte le cose, mentre che sono in moto, più facilmente si muovono, che se di subito uscissero dalla quiete. Et questo in cotal modo si essequisce: Si uscisse fuori dal diritto col piede diritto;

p. 80 §1 & col manco accompagnato da un riverso traversale, s'aiuta à qual si voglia ferita, che venga dalle parti diritte del nemico alla sua declinatione, & poi subentrando col diritto si ferisce la parte più à noi vicina.

p. 80 §2 Cosi in tutte le ferite, che vengono dal lato sinistro del nemico; come sono i riversi di tutte tre le specie, & di quelli composti, uscendo dal diametro del circolo verso le parti manche nemiche con un taglio traversale, aiutaremo le dette linee al suo descenso; & col minor moto à noi più facile feriremo.

p. 80 §3 Nel secondo modo di parare si repugna alla percossa del nemico: il qual viene sconcertato in questa maniera (& questo modo è buono per far cadere la spada di mano al medesimo nemico) subito, che egli fa una delle ferite (ò siano di taglio, ò siano di punta) che vengono dalle parti diritte, si guadagna il terreno col piè diritto seguitato dal manco di moto nostro naturale, mentre che si cercano le parti manche del nemico: & con un mezo mandiritto si taglia quella linea nel suo debole; cioè; nell'ultimo terzo verso la punta della spada; & si risponde con la ferita più commoda.

p. 80 §4 Medesimamente tutte le ferite, che vengono dalla parte manca con la cresciuta del piede manco da un mezo riverso sono segate: il qual fende verso terra;

p. 81 §1 & cosi dalla cresciuta del piede diritto è ferito il nemico.

p. 82 §2 Nel terzo modo di parare si resiste à tutte le ferite col filo diritto; & se bene si potrebbe anchora col falso, nondimeno per esser questa parata debole la tralasciammo: & mettendosi la spada in linea retta si para, & si ferisce facendosi scorrere i fili delle spade: & à tutte le proportioni di linee fatte dalle parti diritte del nemico, si para con la guardia d'entrare col piede manco, overo col diritto avanti: & à quelle, che vengono dalle parti manche, si para vol tando il filo buono in guardia di faccia.

p. 82 §3 Nel quarto, & ultimo, si para à tutte le ferite in guardia di testa; & questo modo è più sicuro del terzo; atteso che non è sottoposto à gl'inganni; & con esso si risponde di quella ferita, ch'è più commoda: avertendosi di ricever la botta nel primo terzo della spada nel più forte d'essa; poi raccogliendo nella parata il manco piede appresso al diritto, potrassi ferire ò di taglio diritto, overo di riverso.

p. 82 §4 Quando ascuno è gagliardo; potrà parare coperto; cioè; con la spada un poco traversata nel forte: & poi ferire con quella ferita, che sarà più commoda alla parata fatta.

p. 81 §2 *E bene, che noi diciamo, che oltre che à questo modo detto (poiche ci viene à proposito) con due altri si può far cadere la spada di mano al nemico.*

p. 81 §3 NEL primo con la spada sola con la Garatusa (che cosi questa viendetta) la qual si fà, & col taglio di ritto, & col riverso: co' quali due tagli si avertirà di torre la spada nemica nel suo ultimo terzo; & cercarassi col pugno della nostra spada più basso di quella punta nemica cõ[n] un groppo di mano fendente al filo della spada del nemico di fargliela cadere: se poi di riverso ciò si vorrà tẽ[n]tare col pugno nell'istesso livello, il quale havrà differenza solo in questo, che la punta della nostra spada mirerà le parti manche nemiche, & userà del riverso à molinello, il quale habbia da fendere il filo falso della spada, facilmente otterremo il nostro fine.

p. 81 §4 Nel secondo si mette la spada nostra sotto quella del nemico, & la punta d'essa s'appoggia sopra il braccio diritto del contrario: poi col nostro pugnale

p. 82 §1 messo in dentro diamo la volta in fuori, facendo l'effetto della lieva: che cosi se gli fa cadere la spada.

p. 82 §5 A tutte queste sorti di proportioni di linee, che sono cosi in stato, come in moto, dovendole noi

p. 83 §1 [supera]re, sarâ necessario, che ci truoviamo col corpo messo in quattro atti; i quali noi chiamaremo guardie: non perche si debba stare in esse per aspettare il nemico: ma affine di stringerlo; & insieme operare l'offesa, & la difesa. Quattro guardie dunque constituireio per esser quattro le nostre ferite; & quattro i modi di operare, le quali mostraremo in disegno, havendo ciascheduna sù la sua ferita.

p. 83 §2 ET la prima sarà questa, la quale si vede qui di sotto; & nella quale, affine che il corpo possa ubbidire alla volontà, habbiamo eletto la prima postura del corpo inventa ta da Policleto, la qual tiene tutto il peso visto nella base manca.

p. 83 §3 Il pie diritto si truova alleggerito dal peso, per esser quello, che hà da entrare nel moto: & si truova il corpo nel primo circolo de i quattro, che forma in moto: accioche in breve tempo possa girare à tutte le bande.

p. 83 §4 L'operatione col piede diritto avanti; come dice Vegetio al vigesimo capitolo; fù costumata da gl'antichi Romani; i quali, quando lanciavano i pili, tenevano il manco avanti: accioche la percossa fosse di maggior vehementia: & quando adoperavano la spada vi tenevano il diritto: accioche i lati

p. 84 §1 fossero coperti, & la spada fosse più vicina al punto nemico.

p. 84 §2 Delle tre proportioni del corpo in questa postura si piglia lo scurzo, ò vogliamo dire la veduta d'una faccia, & meza, per mostrar al nemico minor superficie, che sia possibile: & per non concedere al nemico salvo che un punto: & questo punto, per allontanarlo dalla spada nemica, si piega il corpo avanti col primo suo moto, il quale esso fa con tutto il corpo; come habbiamo detto di sopra; & per conseguente la spada s'avicina al punto contrario.

p. 84 §3 Et perche la spada, mentre che si truova in stato, denota la futura percossa: perciò noi la metteremo in proportione tale, che sia in potẽ[n]za d'effettuar qual si voglia sorte di ferita, senza che dalla postura della mano sia necessitata à fare una sola ferita. Si doverà dunque mettere in stato, onde ritrovandosi il corpo in scurzo, & alzandosi il braccio nel suo naturale stato, la spada, & il braccio formino col corpo ua istessa linea. Talche tirandosi una linea dalla punta estrema della spada, si formi una figura quadrata come dimostrarà la postura.

p. 84 §4 In questa proportione sarà la spada vicina al punto nemico perche voltandosi solo il pugno, & con esso formandosi in moto triangolo equilatero, la spada entrerà in linea retta, mentre che ferirà.

p. 84 §5 Nelle sue ferite descendendo dalla circonferenza al

p. 85 §1 centro di moto naturale d'alto à basso cerchera tutto il corpo, ch'è luogo commune alla ferita: & sarà irreparabile per formar essa nel suo moto un semicircolo, & perche forma il pugno triangolo equilatero, perciò la spada fa tre vedute all'occhio del nemico.

p. 85 §2 Con questa postura l'huomo ferendo si serve della quarta lunghezza, ch'è migliore dell altre tre; come s'è detto: & con una breve di tempo si para, & ferisce, & si fa il scanso della vita.

p. 85 §3 Quando havremo il pugnale, lo terremo in proportione tale, che possa dominare la spada nemica; & che di qui la nostra possa dominare il pugnale del nemico: overo con la mano manca staremo pronti alla difesa.

p. 86 §1 [Illustration of the first position]

p. 87 §1 AVERTIREMO nondimeno, che stando noni in questa postura, potremo entrare nell'altre tre; & ancho dall'una nell'altra di queste, conforme al bisogno, solo facendo moto con una base, & fermã[n]do l'altra come, per essempio; se la spada nemica mirasse il punto nostro diritto, poiche secondo il precetto, rimosso c'haveremo il punto, ove mira detta spada, rimoveremo ancho l'effetto col piede manco uscendo nel secondo circolo, che si fa in moto, di moto traversale, & scaricando il peso del corpo nella base manca, formaremo la seconda guardia.

p. 87 §2 Nella quale poiche il piede diritto hà da caminare alla ferita, egli si truova allegerito dalla parte del suo peso; & la spada naturalmente da se entra in proportione d'imbroccata; & il pugnale pur dominando la spada nemica stà prõ[n]to alla difesa: & cõ[n] l'istessa battuta, & lunghezza di linea si ferisce, come si vede nel la seguente figura.

p. 88 §1 [Illustration of the second position]

p. 89 §1 ET perche; come s'è detto; habbiamo il nostro corpo diviso in due parti eguali dal diametro del circolo; cioè; in parte diritta, & in parte manca: essendo che; come dice Boetio nel libro delle Divisioni; qual si voglia diuisione vuol essere di due membri; se il nemico mettesse la spada retta nel punto nostro manco, portaremo quel punto fuori della spada nemica col diritto piede di moto traversale nel secondo circolo: ilqual facciamo in moto, & co piedi, & col corpo in scurzo formaremo la terza guardia, tenendo il braccio diritto in stato d'imbroccata, & il manco ben disteso; il quale habbia da chiudere fuori la spada nemica: come si vede in questa figura.

p. 90 §1 [Illustration of the third position]

p. 91 §1 STANDO l'huomo in qual si voglia di queste tre guardie; & peraventura, che il nemico mettesse la sua spada in presentia, ma angolare; da quelle solo col moto del piede diritto intrerà nella quarta guardia: se bene questa descrittione, ch'io qui disotto farò d'essa, è particolare da se, senza che l'huomo si truovi posto in alcuna delle dette guardie

p. 91 §2 Si mette il piede diritto appresso al manco nel primo circolo col corpo in billico; & col ritirare adietro il diritto di moto retto per il diametro del circolo; il qual si forma in terra, arrivandosi cosi compassato nella terza circonferenza del terzo circolo col piede diritto; si forma la quarta guardia, con tutto il peso del corpo nella base diritta.

p. 91 §3 Nel ritirarsi il piede diritto si fà un semicircolo con la spada; & questa si posa sotto il braccio mã[n]co trasersata; il qual si tiene disteso, & pronto alla difesa; & in quella battuta, che si fà col diritto, si guadagna col manco un piede di terreno, per ingannare il nemico; & il corpo si tiene sopra la base diritta, per discostarlo dal nemico: & se bene col diritto si ferisce, perche il corpo è quello, che prima si muove alla ferita, nondimeno nõ[n] importa, che il peso sia in quella base. Et la postura è questa, che quì disotto segue.

p. 92 §1 [Illustration of the fourth position]

p. 93 §1 HORA, CHABBIAMO DESCRITTO *i theoremi di quest'arte, ci resta il dimostrare la prattica: ma avanti, che la descriviamo, poiche questa poca digressione, ch'io sono per fare, si confà tanto a proposito nostro, anchorche ciò sia il volere entrare in altra materia: & poiche ella apporterà molto piacere à chi sarà della professione, & di più servirà à maggior intelligenza, hò risoluto di trattare brevemente in generale dell offese, & difese delle fortezze, & di quelle solo, che faranno à questo proposito: & tutto questo poi applicare alla nostra particular facoltà, affine d'imparare à difende re il corpo nostro, & offendere quello del nemico.*

p. 93 §2 POICHE le fortezze, le quali erano fondate ne'siti forti naturalmente; con la esperientia sono riuscite deboli, l'huomo s'è ingegnato con l'arte eleggendo il piano di constituire una fortezza, laquale fosse meno sottoposta à i diffetti, che per il nuovo modo d'offẽ[n]dere sono stati fatti palesi: & questo cõ[n] vario ordine di Bellovardi; cavaglieri, piatteforme, fianchi, cortine, spalle, parapetti, terrapieni, case matte, contramine, fossa, & spalto; & di tant'altre parti, & membra, quante si vede ne'libri del Teti, Tartaglia, &
p. 94 §1 Castriotto, i quali tutti di ciò hanno trattato.

p. 94 §2 Ma la natura havendo formato il corpo dell'huomo di cosi misurata proportione, come s'è detto, hà in esso parimente constituito la fortezza composta del vario ordine, c'habbiamo detto; col quale sono state fatte le fortezze: atteso che egli hà le due mani, come Bellovardi, le quali difendono la cortina, ch'è il corpo; & parimente hà l'altre parti, come si vederà nel seguente trattato.

p. 94 §3 Et per esser le fortezze inespugnabili, converrebbe, che ritenendo sempre tutte l'offese sue, in guisa l'inimico offendessero, che esse in modo alcuno non permettessero l'avicinarsele; ilche fin'ora par che non si sia truovato: percioche l'offesa si fa per linea retta: & con quella via, con la quale quei di dentro offendono i nemici; con quella medesima scoprono il tutto quei di fuori.

p. 94 §4 La fortezza dell'huomo in consequenza di questo hà d'haver queste due conditioni; cioè; l'una è il tenere sempre le sue offese libere, & non sottoposte all'offese del nemico: l'altra il non lasciare avicinare il nemico alla sua distanza. Noi dunque si sforzeremo di tenere la spada, & il pugnale non sotto posti all'offesa, ma fuori di presentia: come s'è detto, parlandosi della prima postura, per questo fine ritruovata dall'Illustrissimo Signor Silvio Piccolomini.

p. 95 §1 Et si come un Bellovardo, quando è priuo delle sue offese, non può soccorrere, & difendere l'altro; ne può medesimamente difendere sa cortina: cosi ogni volta che una delle mani, che sono quasi Bellovardi della cortina dell'huomo, sarà in maniera impedita, che non possa essequire la sua offesa, l'altra resterà sottoposta al nemico; al qual diffetto rimediaremo, se le terremo libere (come s'è detto,) & pronte all'officio loro.

p. 95 §2 Il non permettere al nemico, che s'avicini alla fossa, è tenuto da tutti i valenti soldati la miglior difesa, che si faccia: perche perduta la fossa, la fortezza si dice esser mezo guadagnata: & perciò è parer de' valent'huomini, che si debba mettere ogni sforzo in difendere la strada coperta; la quale, quando sia ben fatta, facilmente si difendera, & dalle mura con l'archibuseria; & col mantenere in essa buona soldaria; la quale à tempo, & luoco faccia delle sortite adosso al nemico.

p. 95 §3 Da quest'utile di mantenersi il nemico lontano, doviamo di continuo oprare in modo, che'I nostro contrario non s'accosti mai à mettere il piede sopra la circonferenza del secõ[n]do circolo, il quale formiamo in stato (come habbiamo di sopra detto) che l'orlo del fosso della nostra fortezza, tirando buone imbroccate aventate dalle mura; cioè; dalla postura, senza aprire il compasso; & con sortite,

p. 96 §1 aprẽ[n]do le nostre colonne, & entrando nella quarta cir conferenza; la quale formiamo in moto, compassati in passo sforzato, (come s'è detto) accioche con maggior lunghezza potiamò prevenire il nemico.

p. 96 §2 Le sortite si fanno con occasione di pioggia; overo d'haver tenuto il nemico affaticato in continue armi; & nell'hora del desinare; ch'è la più propria: quero secondo, che giudica meglio colui, che si truova nel fatto.

p. 96 §3 Parimente noi con occasione, che'l nemico si truoverà affaticato in moto, lo feriremo sortendo nella quarta circonferenza, perche mentre che sarà in moto non potrà rispondere con la ferita.

p. 96 §4 Quando poi s'effettua detta sortita, s'elegge parte della buona soldaria ben armata; & con secretezza si conduce nella strada coperta: & all'improviso s'assalta il nemico: & acciò che la detta soldaria nella ritirata sia sicura, sarmano le mura d'archibuseria; laquale habbia da difenderla.

p. 96 §5 Nell'istesso modo noi delle ferite eleggeremo la migliore, & all'improviso feriremo il nemico; poiche egli, vedendoci tanto lontano dal suo corpo, non aspetta tal ferita; ma per poterci ritirar sicuri lo feriremo d'un riverso traversale nella ritirata, che noi faremo alla strada coperta.

p. 96 §6 Ma l'offesa, che di giorno non si può far contra le fortezze, più facilmente di notte si essequisce,

p. 97 §1 [guadagnan]dosi lo spalto; & per conseguente s'arriva alla fossa; & perche la vera offesa è quella, che si fa, quando siamo sicuri di non esser offesi, si debbono tirar le corde, & far le trincee: poiche sicuri con questa via, & coperti s'aviciniamo alla fossa.

p. 97 §2 Cosi noi, mentre che offenderemo il nostro contrario, sempre faremo col nostro pugnale un semicircolo in moto, il quale ci difenda il nostro corpo.

p. 97 §3 La trincea dominerà l'angolo del Bellovardo in dentro, accioche non sia imboccata: deve anchora esser breve, accioche il nemico, scoperta la volontà del suo contrario, non habbia tempo da provedersi alla difesa.

p. 97 §4 Cosi se il nostro nemico cercasse coperto di dominarci la spada, ò il pugnale, (poiche le mani, che sono i nostri Bellovardi, sono mobili) non lo permetteremo: anzi di continuo dominaremo in dentro la spada, & il pugnale del nostro contrario, che in tal modo il nostro corpo non sarà imboccato; cioe; ferito.

p. 98 §1 *Presupposto, che il nostro contrario habbia fatto le trincee; & che si sia accostato alla fossa, volendo con esse entrare nella fortezza, converrà che si faccia la strada con l'offesa.*

p. 98 §2 ET in tre modi si offendono le fortezze, overo con le sotterranee occulte; cioè; con cave, & mine; overo con le palesi sopra terra; cioè; con la batteria, con la zappa, & pala, & con fornelli; overo con l'aere; cioè; quando con cavaglieri s'alza in modo, che quei di dentro non possono apparire alla difesa.

p. 98 §3 Parimente in tre modi potrà esser tentata la nostra fortezza: con le ferite basse, con quelle à mezo, & con l'altre per di sopra.

p. 98 §4 Quando s'offende le fortezze con le cave & mine, all'hora il difensore cerca di riconoscere il luoco, ilche fa con varii instrumenti: percioche piccando quei di sotto, s'hà con tali instrumenti segno di quel moto; & assicurato il difensore del luoco, ò quella parte taglia, distaccandola da tutto il corpo della fortezza; ò con una istessa cava riscontra l'altra, per far vana la mina, operando ch'ella eshali; ò cerca di guastarla con acqua, ò con fuoco.

p. 99 §1 Nell'istesso modo, quando saremo offesi con le linee basse, come quando l'inimico fa l'inquartata, le rincontraremo piegandosi nel cinto; & non le lasciaremo capitar dentro al recinto del nostro corpo.

p. 99 §2 L'offese palese sopra la terra, principalmente si fanno cõ[n] l'artiglieria: & con essa si fa la breccia, & si levano le difese, perilche si guadagna la fossa, la quale guadagnata si potrà poi far l'offesa con la zappa, & pala, & co'fornelli.

p. 99 §3 Però l'artiglieria si pianta più vicino, che sia possibile, se il sito lo concede al luoco, nel quale si deue battere: perche ogni peso di moto violento eietto per l'aere, quanto più sarà vicino alla possanza movente, ferendo in un corpo, che resista; tanto maggior effetto farà: & quell'effetto è importantissimo, per ottenere il fine in poco tempo: il quale è pretiosissimo nell'offendere le fortezze: oltre che si fa minor spesa.

p. 99 §4 Più d'appresso dunque, che fia possibile, ferireno; cioè; quando il nemico sarà tanto lontano da noi, che siamo sicuri di aggiungerlo col nostro aprirci in forza: perche se da lontano incominciaremo l'offesa, perderemo il tempo, mostraremo viltà, & scoprendoci daremo occasione al nemico di ferirci, & quando fossimo un puoco lontani cresceremo col manco piede prima, & poi col diritto faremo la ferita.

p. 100 §1 L'effetto maggiore, che faccia l'artiglieria nella sua percossa, è quando la palla ferisce la piana superficie del muro in angolo retto: perche il diametro della palla opera tutto nell'offesa: che quando la palla colpisse obliquamente parte del diametro tocca: & parte va a vuoto.

p. 100 §2 Nell'istesso modo la ferita si deve fare in angolo retto; cioè; operaremo che la spada nella sua ferita formi con la piana superficie del corpo angolo retto: che in tal proportione la spada contraria diviene obliqua al corpo nostro: oltre che, quando le ferite sono oblique, con ogni minimo scanso del corpo si fuggono, & si rendono inutili: & sempre feriremo in angolo retto, se seruaremo le regole dette di sopra.

p. 100 §3 Il luoco, dove si deue fare la batteria, è la faccia d'uno de Bellovardi, atteso che levate le difese dell'altro, che quella parte difendeva, & quelle d'alto, si camina sicuramente all'assalto.

p. 100 §4 Parimente noi batteremo il Bellovardo nemico, che sarà quella mano, che egli ci porgerà avanti: & per conseguente terremo le nostre lontane dall'offesa.

p. 100 §5 Vogliono alcuni, che si debba battere la cortina, per esser ella più debole del fianco, & della faccia del Bellovardo; & questo per l'ampiezza sua, & per lo spatio, ond'egli hà di far la ritirata: per beneficio

p. 101 §1 del quale si può stare à cavagliero à gli assalitori: ilche non aviene battendosi la cortina; non v'essendo luoco da ritirarsi, salvo che nel piano: oltre che si sa più ampia la via. Ma non considerano costoro, che s'andarebbe in mezo à cinque offese, le qua li sono quattro per fianco, & una per fronte.

p. 101 §2 Però si cerca sempre d'offendere le parti, che sono mẽ[n]difese: & per questo sempre feriremo la parte del corpo, la qual porgerà più in fuori; & che à noi sarà piùi vicina: percioche ella serà la più debole, come pure è nelle fortezze, & mancherà di difesa.

p. 101 §3 Poiche veggiamo, che la più difesa parte è la cortina: atteso che i Bellovardi si fanno per difenderla: & di questa più quella parte, che fa angolo col fianco: perciò sempre concederemo al nemico il punto della cortina nostra, ch'è il fianco diritto: accioche potiamo truovarsi in maggior difesa; & necessitiamo il nemico à sottoporsi à molte offese.

p. 101 §4 Onde, quando il nemico vorrà ferire in detto luoco, con l'offese orizontali; come con le stoccate; col Bellovardo manco si batterà detta linea; & col diritto si ferirà.

p. 101 §5 Et se alcuno volesse battere il Bellovardo manco, egli sarà necessitato à passare con la prospettiva del suo corpo nella linea retta imaginata; nella quale scorti nerà il Bellovardo diritto; & all'hora sarà occasione di ferirlo.

p. 102 §1 Similmente, se si volesse attaccare al diritto, & con l'istesso, & col manco l'offenderemo.

p. 102 §2 SI fanno l'offese aeree, quando si compone una montagna per di fuori: & con quella si stà à cavagliero à quei di dentro; i quali non possono apparire alla difesa: perche si batte per cortina; & si scuopre dentro in ogni parte, & quest offesa è la maggior, che si faccia.

p. 102 §3 Onde noi sempre con la nostra spada staremo à cavagliero al nemico, & quello offenderemo con l'imbroccate aventate da alto à basso.

p. 102 §4 Quelli di dentro si possono difendere anchor essi con alzarsi, se fia à loro possibile; & col far contrabatteria, overo possono far delle traverse, overo possono risoluersi in tempo; & di quà cercar d'ouviare, & impedire la fabrica al nemico; overo pur risolvendosi in tempo possono caminarvi sotto con una mina.

p. 102 §5 Onde noi, conforme all'alzamento, che farà il nemico con la sua spada, alzaremo il nostro pugnale: il quale s'hà da truovare sempre à cavagliero alla spada contraria.

p. 102 §6 Ma ritornando alla batteria, dico, che presupposto, che con essa si sia fatto la breccia, si mã[n]da un

p. 103 §1 [solda]to prattico à riconoscerla; il quale deue considerare, se la salita è facile; se il camino da farsi nel fosso, è sicuro; & principalmente se nelli fianchi fosse qualche casamatta: & di dentro deve vedere se il nemico havrà fatto ritirata, & in qual modo si truovino à star dentro tutti i luochi: & di qui conforme al riporto del medesimo soldato, delibera il capitano.

p. 103 §2 Nell'istesso modo, fatta che noi havremo la ferita, cer cheremo di riconoscere quello, che si faccia il nemico: perche per l'essere egli fatto iracondo dalla percossa, dovremo star sopra di noi, & avertiti: & più pronti alla seconda ferita, che alla prima.

p. 103 §3 Le ritirate si fanno in quattro modi; i quali tutti nondimeno fanno il medesimo effetto. Nel primo si tira una fossa in linea retta, per tutto lo spatio del muro rotto: & si fanno le casematte nelli fianchi d'essa fossa di legname; & della terra che si cava, ma più di legni grossi messi in punta: si fa etiandio il terrapieno buono, & eminente, & col parapetto di gabbionate: Nel secondo si fanno à meza luna: Nel terzo ad angolo: Nel quarto à tenaglia. Et si deve lasciar quel manco terreno, che sia possibile, al nemico: accioche egli non habbia sito, da condurvi la sua artiglieria, & d'alloggiarvi sopra.

p. 103 §4 Queste ritirate sono molto migliori, quando vi sia sito da poterle fare, che non e il stoppar il vacuo

p. 104 §1 della breccia: perche non si può lauorare se non di notte; & questo ancho alle volte si può far malamente: & perciò il riparo sarà debole.

p. 104 §2 Cosi, fatto che noi haveremo la ferita, faremo la ritirata con tre passi, sempre tenendo il corpo in prospettiva: accioche il nemico non possa ferir noi, mentre che staremo fermi nella distanza, che tal nostra ritirata sarà à meza luna.

p. 104 §3 Onde, quando le ritirate sono fatte, come si conviene, sono molto difficili da guadagnarsi: perche, qual si voglia proportione di linea, che entri in una circonferenza, viene imboccata da detta circonferenza.

p. 104 §4 Però noi staremo sempre in stato di ritirata col corpo in semicircolo, necessitando il nemico à venire in mezo della nostra ritirata; la quale sarà la spada, & il pugnale posti nella nostra guardia: perche, qual si voglia sorte di ferita, che capiti in mezo delle nostre armi, sarà imboccata dal pugnale; & sarà ferito il nemico dalla nostra spada.

p. 104 §5 Et quando si vuole entrare in dette ritirate, si camina con la trincea alla Francese: la quale è difficilissima, & di molto stento: peroche cavando bisogna buttarsi il terreno dalle bande per coprirsi: & per fronte si mettono fascinoni; overo tavoloni: ma il difensore aggiustandovi un pezzo d'artiglieria, fa grandissimo danno. Oltre, ch esso medesimamẽ[n]te

p. 105 §1 camina all'incontro; & con molti ordegni cerca di ributtare il nemico.

p. 105 §2 Parimente, se il nostro nemico per disperato volesse entrare nella nostra ritirata, coprendosi col pugnale, & con la spada in croce; noi ributtandolo lo incontraremo con la ferita risoluta.

p. 105 §3 Ma il meglio, che si faccia, è di cogliere il nemico per le spalle, affine di rendergli inutile tutta quella fortezza fatta, battendo pur di nuovo alle spalle della ritirata; overo caminando coperto con la zappa, & pala, à che l'esperto capitano di nuovo rimedia rinchiudendo il luoco battuto con un altro semicircolo.

p. 105 §4 In questa guisa volendo il nemico ferirci alle spalle; noi fattine accorti con un semicircolo lo rinchiuderemo in mezo delle nostre forze; & in quei tempi lo feriremo.

p. 105 §5 Onde, con questo modo d'offendere, & di difendere, si và procedendo per fin tanto, che si riducono all'estremo; & quiui il difensore tagliando distacca la parte offesa, & resta in isola, necessitando il nemico à pensare à nuove offese.

p. 105 §6 Ridotti, che saremo a questo fine, che sarà quando ci saremo lasciati ingannare dal nemico, il quale sarà arrivato alla nostra distantià, perderemo un passo di terreno; & formaremo nuovo circolo: & cosi daremo causa al nemico di ferirci: al qual

p. 106 §1 pericolo con le nostre solite parate, & ferite rimediaremo.

p. 106 §2 Resta, che per ultimare questo trattato, diciamo dell'assalto, il quale si può dar con valore, se con diligenza arrivando col campo il generale, & con prestezza, (havendosi prima riconosciuto il luogo più debole) fa piantar i pezzi per levare le difese, & i canoni per far la batteria: accioche il difensore soprassaltato perda l'ardire: & non habbia tempo di fare le difese; & insieme fa far gli approcchi.

p. 106 §3 Con questo modo procedendosi, riconosciuto il nemico in qual stato egli sia; & in qual proportione di linea egli tenga la sua spada; & il luoco più debole; cioè; quella parte del suo corpo, che sarà à noi più vicina con prestezza, & risolutione lo accommetteremo, & lo feriremo.

p. 106 §4 Quando poi si dà l'assalto, il generalissimo dispone tutt'il suo essercitò in battaglia nelle sue piazze d'armi, & mette la cavaglieria alle venute, per le quali potesse il nemico offendere il suo essercito; & quelli, à quali è commesso, che siano i primi à dar l'assalto, entrano nella fossa per la scanatura d'essa, guidati da iloro capitani, dopò i quali vanno gli alfieri con l'insegne.

p. 106 §5 Ordina poi à gli altri, i quali hanno da soccorrere, & rinfrescare l'assalto, che stiano pronti: de quali hà cura il Mastro di campo di quello reggimento.

p. 107 §1 Di parte dell'artiglieria poi esso dispone, che tiri à gli ultimi termini del muro rotto: accioche offenda i difensori, i quali hanno da star quiui di dietro, per ferire di fianco gl'assalitori: & i pezzi, che son posti per levar le difese, frequentano il tirare alli fianchi.

p. 107 §2 Parimente dal Trincerone si frequenta lo sparare la moschetteria, & archibugieria, della quale tutto il Trincerone è armato.

p. 107 §3 Con questo modo disposte l'offese, le quali offendendo difendono, si spara la camerata dell'artiglieria: & sotto quel favore, dandosi nelli tamburri, si rimette all'assalto: il quale poi vien rinfrescato dalle genti già commandate; si come di mano in mano, secondo il bisogno, si comanda loro; le quali condotte dal Sergente maggiore fino al bordo del fosso, animosamente assaltano la fortezza.

p. 107 §4 Il difensore stà in battaglia con la sua gente; & con parte di quella ributta gli assalitori: & oltre di ciò fa ancho questo con l'artiglieria, la quale a vicenda si spara, & con più altre cose s'aiuta, le quali taccio per brevità; ma la miglior difesa, che sia, è di far continuare le maniche d'archibugieri, le quali à vicenda in caraccollo seguitano di sparare.

p. 107 §5 Manda oltra di ciò solderia nel fosso, affine che possa offendere gli assalitori per le spalle: alla quale s'oppongono i soldati, che sono nelle trincee, & gli fanno ritirare.

p. 108 §1 A tutto questo modo dunque la volontà nostra, capitano generale del nostro essercito, & della nostra fortezza, volendo assalire, ciò farà con diligenza, & risolutione; & disporrà l'essercito suo in battaglia, con ordinare al senso esteriore della visuale, che debba guardare la venuta della linea del nemico nel suo estremo: atteso che quella deve ferire il corpo, & non la mano: come altri vogliono.

p. 108 §2 Et oltre a questo, per poco moto che faccia la mano, la punta della spada ne fa moto molto maggiore; & perciò se noi guardaremo alla mano, perderemò il tempo di ferire: perche non vederemo in tempo l'angolo constituito dalla linea imaginata, & dalla linea della spada.

p. 108 §3 Ordinarà parimente alla potenza appetitiva concupiscibile, & irascibile, che con essa s'opponghino alle cose, che impediscono il nostro fine: & che commandino alla potenza motiva, che debba stare all'erta, per dar il moto alle nostre membra: le quali havranno da stare unite: accioche bisognando, pos sino dar l'assalto, & ubbidire alla volontà.

p. 108 §4 Parte delle quali membra anderà all'assalto; & parte difendera quelle, che faranno l'offesa: come farà il pugnale, quando difende la parte diritta del nostro corpo; ch'e quella, che và all'offesa; & cosi facendosi, otteniremo il fine nostro, che sarà la vittoria.

p. 108 §5 *Il fine della Theorica.*

p. 109 §1

DELLA PRATTICA

p. 109 §2 *La nostra prattica, si riduce à poche cose: perche con la nostra operatione neghiamo tutti i principii delle varie offese, che si possono fare.*

p. 109 §3 SUBITO dunque, che si scuoprirà il nemico da lontano, si lasciaremo cadere la cappa dalla spalla diritta; & con la mano manca per di dietro si spingera il pugnale avanti su'l fianco diritto. Poi prima si trarà fuori del fuodero la spada; & appresso il pugnale: & in un tempo si stringeranno le pugna, & i denti; & s'allargheranno gli occhi mostrando nel viso fierezza.

p. 109 §4 Doppo questo co' l'armi basse spinte fuori del corpo, & dispostamente co passi naturali si caminerà; & prima che s'arrivi alla distantia riconosceremo il nemico; & staremo avertiti con qual piede egli si presenterà avanti, & con qual arme: perche s'egli terrà il pugnale avanti questo farà per difendersi; se la spada, per offendere; tuttavia faccia esso in qualsi voglia modo, sempre saremo noi quelli, che accometteremo, & che gli daremo il moto.

p. 109 §5 Et presupposto che il nemico si truovi in stato con la

p. 110 §1 spada in atto nella linea retta, ma dominabile dal nostro pugnale; come in guardia terza, & quarta; si presentaremo col corpo in prospettiva, & nel stretto intraremo nella prima postura, nella quale servandosi le regole dette, potremo con una breve ferire; & nell'istesso tempo potremo battere la spada nemica, & distorla da quella linea, aiutandola alla sua declinatione.

p. 110 §2 Perche, se consideriamo le quattro cause da noi dette disopra; delle quali la prima è la postura del corpo: per quella siamo sicuri, che quel corpo in moto puo venir rettamente: & la proportione della spada, causa formale, ci mostra la ferita per linea retta. La causa materiale ci palesa, che ogn'altra sorte di moto, che il nemico faccia ò col corpo,ò col braccio, ò sia di spingimento, ò di ritiramento, ò di portamento, ò di giramento; sarà con maggior tempo fatto, che non sarebbe il moto, qual indica la postura; la causa finale, che è il punto estremo di tal linea, hora dal pugnale è rimossa, perche quella battendo la distoglie dalla sua proportione.

p. 110 §3 Et affine, che l'huomo non si confonda, & pigli errore; doviamo sapere, che tutte le regole, c'habbiamo detto, servono cosi al giuoco di spada, & pugnale, come di spada sola: ma hanno solo differenza in questo; che quando noi habbiamo il pugnale, & che combattiamo contra le linee rette, le quali

p. 111 §1 [so]no in presentia, ma sottoposte al nostro pugnale, all'hora non occorre, che portiamo fuori i punti, i quali sono le cause finali di coteste linee; poiche col pugnale quelle battendo rimoviamo l'effetto della sua ferita.

p. 111 §2 Ma con la spada sola, la quale è priua di tale difesa, quando contra queste proportioni di linee, combatteremo sempre ferendo, ciò faremo col piede manco, bisognandoci rimuovere il punto coi moto del corpo; poiche non habbiamo il pugnale: & con la mano manca chiuderemo il ritorno alla spada del nemico, come dimostrano le due figure piccole, che sono in questa dimostratione prima, che qui di sotto seguirà.

p. 111 §3 Et contra alle linee rette in presentia alte, & con l'arme accompagnate: & con la spada sola, operaremo in un modo istesso.

p. 111 §4 Combattendosi poi contra alle linee oblique, terremo l'istesso ordine: & solo differiremo in questo, che con la spada sola, quando feriremo di necessità; cioè; nel tempo mentre che il nemico ferira noi, ciò faremo con la linea retta, col corpo in profilo.

p. 111 §5 Et ritornando alla dichiaratione della Prattica, dico, che ritruovandosi il nemico in terza, ò vero quarta guardia, ò sia con le gambe aperte, ò strette, si presentaremo contra à quello in prospettiva, & mettendoci in prima guardia col manco piede nel

p. 112 §1 diametro del circolo, subito che la nostra visuale, passando per il forte del nostro pugnale ferirà nel debole della spada nemica; con risolutione feriremo avanti il tempo alle parti diritte: come notifica questa figura, che segue, nella quale; si come in tutte l'altre, è solo la guardia con la sua ferita, che si fà di volontà avanti il tempo: & vi è anchora certa linea tirata dal pugno della nostra postura, la quale significa la ferità di necessità in tempo; cioè; il modo di ferire, quando il nemico ferisce; in oltre quella linea, la quale sortisce dal pugno del contrario, denota la prima postura della spada dell'istesso nemico.

p. 113 §1 [Image of the first posture being used against an opponent.]

p. 114 §1 Non restarò di dire, che se il nemico nostro fosse in stato con la linea retta in presentia, ma ben distesa, (come fanno molti per tenersi il nemico lontano) all'hora quella spada parata batteremo col pugnale, con quello descrivendo un semicircolo in fuori, accioche si possa cavare dal nostro corpo maggior lunghezza nella ferita.

p. 114 §2 OGNI volta che il contrario nostro fusse posto in prima, overo in seconda guardia, ò in guardia di faccia, ò d'entrare, che cosi in questi atti havrà la spada in atto nella linea retta; ma alta, contra à quello operaremo in questo modo; si presentaremo col corpo in prospettiva nel primo circolo, & mettendo il piede diritto nel diametro del circolo, di nostra volontà opporremo il punto diritto, & col manco usciremo di moto traversale, nel secondo circolo, il qual formia mo in moto, che cosi faremo; la seconda guardia è ò in tempo, ò avanti il tempo, feriremo battendo col nostro pugnale.

p. 114 §3 Et tre cose potrebbe questo nostro contrario operare; la prima nell'appresentarsi, che noi facciamo, potrebbe finire di caminare quella linea ferendo, ma noi col tempo premeditato feriremo; la seconda,

p. 115 §1 potrebbe costui nel moto, che noi facciamo, quando uscimmo col punto ferirci: ilche non per scientia; perche il moto nostro è di breve tempo: di maniera, che prima havremo operato, che egli l'habbia veduto. Nondimeno presupposto, che ci succedesse, già si fà il scanso; & s'hà la mano, & il pugnale alla venuta; la terza è, che si potrebbe costui ritirare quando noi battiamo la battuta in prospettiva: la qual sua ritirata non potrà esser di tanta quantità, di quanta sarà la nostra cresciuta: poiche noi ci truoviamo uniti; onde in questo caso feriremo al nostro solito d'imbroccata aventata.

p. 115 §2 Se avanti che l'huomo arriva nello stretto, il nemico mutasse guardia, s'avertirà il modo, col qual si metterà; & conforme alle regole dette cosi si procederà contra esso.

p. 115 §3 Hò voluto dire queste poche ragioni per maggior intelligenza, se bene non occorreva, perche tutte le figure, che qui di sotto si metteranno, sono fatte fondate in tutte le regole dette, & questa, che hora segue, dimostra l'effetto della seconda guardia, & quella linea, che esce dalla spalla del nemico, significa la proportione della spada del contrario, nella quale si suppone, ch'ella fosse.

p. 116 §1 [Image of the second posture being used against an opponent.]

p. 117 §1 SE nell'istesse guardie si truovasse posto il nemico: ma che ci tenesse la spada fissa nel punto nostro manco: noi subito metteressimo il piede manco nel diametro al diritto di quello, che tenesse il nemico avanti; & poi uscendo col diritto di moto traversale nel secondo circolo formaremo la terza guardia; & d'alto à basso feriremo d'imbroccata, lasciando operare la difesa al braccio manco naturalmente, come se fosse d'un pezzo solo, cosi con la mano, come col pugnale.

p. 117 §2 Potrebbe il nemico, quando noi andiamo alla ferita, cambiare per di sotto la sua spada; ma solo riportandosi il peso del corpo nella base manca, & con quella solo facendosi un picciol moto, si batterà quella spada, & si farà la medesima ferita.

p. 117 §3 Non starò à replicare tutte le ragioni, perche quello, che s'è detto della prima guardia, s'intende ancho detto di tutte le linee rette : basta che queste figure dimostreranno l'effetto, come fà questa quì disotto, laqual dimostra come si faccia ferendosi avanti il tempo, & nel tempo.

p. 118 §1 [Image of Ghisliero's Third Position being used]

p. 119 §1 MA quando il nemico havrà la spada in presentia, ma obliqua sotto il centro del nostro corpo, all'hora s'appresentaremo nello stato uniti, & col corpo ben curvato in prima guardia, opponendo il pugnale, corpo resistente alla linea obliqua, accioche non monti, feriremo con la ferita solita, eletta per la migliore, che si possi fare.

p. 119 §2 Ma è d'avertire, che tutto il pericolo e , quando noi battiamo la battuta nell'accostarsi: & perciò col tempo premeditato (non potendo quella linea obliqua con breue tempo far altro, che nella ferita montare in linea retta) sopraporremo il nostro pugnale alla linea imaginata, alla quale deve capitare detta spada, & questa linea imaginata appare qui sotto in questa figura.

p. 120 §1 [Illustration of the first position being used against someone in a low 3rd]

p. 121 §1 SE il nemico tenesse la sua spada obliqua in presentia nel secondo modo, noi subito dopò che si saremo presentati nel diametro, col corpo in prospettiva; intraremò nella minor portione del circolo della distanza, ilquale è segnato con i punti, si come è l'altro, picciolo, che pur esso formiamo in stato: & mettendo i piedi sopra la corda del detto circolo maggiore formaremo la prima guardia: che cosi guadagnando la distantia, (caso che egli ferisce,) parimente noi feriremo, mettendo la nostra spada nella diagonale, & il nostro pugnale nella linea imaginata, nella quale deve venire quella spada obliqua, come dimostra molto bene questa figura, nella quale si vede l'effetto, che fa la nostra ferita; & l'effetto che fà la ferita del nemico.

p. 122 §1 [Image of a fencer using the first position against someone in 2nd]

p. 123 §1 POTREBBE il contrario nostro truovarsi con la spada obliqua al corpo nostro; nel terzo modo; come di sopras'è detto; cioè; angolare, pero avertiremo se l'estremo d'essa sarà al diritto delle parti nostre manche; & subito in arrivando alla distantia si metteremo nella quarta guardia; & feriremò con la maggior lunghezza del corpo; movendo prima lui; & poi seguitando il moto col piede diritto segaremo quella linea; & l'aiutaremo col pugnale alla sua circonferenza: sicuri, che stante in quest angolo, la spada potrà farsi linea retta, alla quale noi col portar il peso del corpo nella base manca, soprasteremo ferendo di prima guardia: & se quando si appresenteremo la medesima linea angolare si facesse retta, ritornaremo nel primo circolo in prima guardia; & operaremo, come habbiamo detto, trattandosi della prima guardia.

p. 123 §2 Ma se stando la spada in angolo, ella fosse posta al diritto delle parti nostre diritte, che in tal postura vederessimo il corpo suo, & non la punta, si appresentaremo in prima guardia, operando contr'essa, come se fosse linea retta.

p. 123 §3 La figura, che qui di sotto segue mostra molto bene l'effetto del nemico, & l'effetto della nostra postura.

p. 124 §1 [Image of the fourth position being used against an opponent in an angled 3rd]

p. 125 §1 *Hor c'habbiamo detto delle linee oblique in presentia, ci resta per venir al fine di questo trattato, che di quelle ragioniamo, le quali si truoveranno fuori di presentia, potendo il nemico ritruovarsi posto in esse in due modi, ò col diritto piede avanti, ò col manco.*

p. 125 §2 HAVENDO noi per fine di ferire il nostro nemico, ogni volta che egli terrà la sua spada adietro, l'istesso ci facilita la nostra operatione; & terremo per regola d'allargarsi dalla spada nemica altro tanto, quanto quella sia lontan'a da noi, & cosi facendo cercaremo il corpo nemico, & tenendo il corpo in scurzo usciremo nella minor portione del circolo: & stando nella corda di detto circolo feriremo avanti il tempo, & nel tempo nell'istesso modo, come appare qui in questa figura che seguita.

p. 126 §1 [Image of the first position being used against an opponent in a low refused 3rd]

p. 127 §1 FINALMENTE ritruovandosi il nemico in stato col piede manco avanti, egli in questa postura nõ[n] concede se non un punto della sua superficie: tuttavia quando vorrà ferire, bisogna che passi cõ[n] tutta la prospettiva del corpo nel la ferita: & questo farà col moto maggiore, che si possa fare: il quale sarà descritto col tempo d'una massima; che vale otto battute. Noi dunque entraremo nella distantia, mettendoci nella minor portione in prima guardia con la vita: la qual sia più tosto vicina alla proportione del profilo, che altrimenti; & terremo il pugnale, alto se'l nemico sarà in guardia di falcone.

p. 127 §2 Mentre che noi operaremo nel detto modo, daremo il moto al nemico: portaremo fuori il punto: guadagnaremo la distantia: avicineremo la nostra spada al nemico: & col pugnale pronti, il qual terremo in modo, c'habbia da dominare la spada, feriremo il nemico ò avanti il tempo, ò nel tẽ[m]po: & per che meglio s'intenda questo Capitolo, l'hò diviso, & hò fatto due demostrationi; delle quali la prima mostra, che quã[n]do si ferisce di volontà, allhora nõ[n] si serviamo più della nostra ferita solita; ma quella col corpo in profilo operiamo, come si vede nella seguente figura: & si ferisce alle parti manche.

p. 128 §1 [Image of the first position being used against someone in a refused high 1st]

p. 129 §1 OGNI volta, che si dà il moto à questi, che si truovono in stato col piede manco avanti: forzosa mente questi si vogliono accomodare; perche noi levando il punto si copriamo dietro alla linea del corpo loro: ond'essi non scuoprendo luoco, dove poter ferire, si muovono. Pero sarà tempo di ferirli in quel moto, ò sia del diritto piede, ò del manco.

p. 129 §2 Ma se vorranno ferire, noi con la nostra solita parata, & ferita feriremo nella parte diritta, che è quella, che camina alla ferita: come si vede nella seguente.
figura.

✣

p. 130 §1 [Image of the first position being used against someone in a low refused 2nd]

p. 131 §1 TUTTE queste cose, che si sono dette, si dovranno sapere mettere in pruova, secondo il bisogno: perche potendo il nemico (come per essempio) mettersi in guardia d'entrare, & poi abbassare quella linea; & oltre di questo liberarla, & metterla nella linea obliqua; parimente la potrà mettere da una in un'altra proportione. Però osservã[n]dosi le regole dette, davna guardia entraremo nell'altre con grandissima facilità.

p. 131 §2 Siamo fin qui da queste considerationi fatti chiari, che il mettersi in guardia è molto nocivo. Onde segue, che noi in ogni nostra attione non doviamo mai dare inditio del nostro volere: ma haveremo con ogni risolutione da mettere il pugnale nella linea retta imaginata; & con una battuta feriremo ò di punta, la qual sarà d'imbroccata aventata; ò di taglio fendente con tutta la nostra forza, & lunghezza, necessitando il nemico al parare: & subito ricolpiremo con un riverso traversale aiutato dalla forza del pugnale, doppo il qual incominciaremo di nuovo la prima ferita; & questo riverso faremo stanti nella prima apertura in passo di forza, che havremo fatto nella ferita: ma se'l nemico per la prima nostra ferita perderà terreno, faremo questo riverso crescendo col piede manco.

p. 132 §1 Et questo modo di fare si essequirà, avertendo che l'arrivar alla distanza, & il ferire sia tutt'uno.

p. 132 §2 Da questo rimettere risolutamente, & con prestezza, si rende il nemico occupato alla difesa: & di più perde il consiglio, & l'animo; & per consequenza la forza. Et di qui si vede chiaramente, ch'egli si metterà in difesa; overo vserà qualche sorte di ferita involuntaria: allequali ferite fatte per necessità; poiche noi saremo quelli, che gli sforzeremo ad operare contra sua voglia; facilmente resisteremo: & questo tanto più, quanto più elle si veggono esser deboli, &imperfette.

✣

p. 133 §1 AVERTIMENTI DI SPADA, ET PUGNALE.

p. 133 §2 *Volendo noi metterci nella nostra prima guardia, potrebbe il contrario in cinque modi operare contra la detta postura: ma tutto farà in Vano, se noi operaremo secondo la ragione, et la scientia del giuoco.*

p. 133 §3 NEL primo modo cercherà di accostarsi nella distantia; & con velocità s'aventerà alla spada cõ[n] la sua al traverso, per ferire col pugnale. Ma à questo si rimedia col ferire il nemico in moto, avanti che arrivi alla distantia.

p. 133 §4 Oltra di questo con ogni poco moto alzando la nostra spada, la metteremo nella ferita in linea retta. Di più passaremo col manco piede, & feriremo col pugnale.

p. 133 §5 Nel secondo modo, studiarà il contrario d'andare alla spada, per ferire di ponta riversa: ma à questo si rimedia facendo centro col diritto piede: perche cosi non se li permetterà l'avicinarsi: avenga che in quei suoi moti lo potremo ferire; & presupposto, che egli cavasse la ferita riversa, noi cacciaremo la

p. 134 §1 nostra spada in linea retta, col corpo in profilo, che cosi segaremo quella linea.

p. 134 §2 Nel terzo modo, con la spada bassa giostrando, il contrario farà l'inquartata: & perciò noi si moveremo giostrando per di sopra; & col pugnale riteneremo la spada che non monti.

p. 134 §3 Nel quarto modo farà il contrario montare la spada per mezo, non per ferire: ma per cacciare la punta riversa, à che si rimedia con l'istesso modo detto.

p. 134 §4 Nel quinto modo, metterà il contrario la punta della spada nel pugno, per impedirla: accioche non descenda. A questo non consentiremo, se subito, che la spada nemica ci mostra il corpo, feriremo: ma presupposto, che già la spada sia messa nel nostro pugno, faremo un poco di segno di ritirata, solo col pugno della spada: perche s'egli rimetterà la spada in quel moto lo feriremo.

p. 134 §5 Se nell'appresentarsi l'huonio si fosse ingannato, ò si fosse sconcertato, deve ritirarsi tirando un riverso circolare, il quale fenda la terra: & poi riconosciutosi si debbe rimettere.

p. 134 §6 Quando si sente impedita la spada, si deve cedere à quella forza; & ferire di quel moto, che l'istessa forza fa fare, & questo modo di operare nell'armi è perfettissimo.

p. 134 §7 Se il nemico contrapassa, bisogna che noi parimente contrapassiamo: & di quà il secõ[n]do hà avantaggio.

p. 135 §1 Potendosi ferire con quattro dita di spada, non bisogna voler soprafare: perche la prima ferita va fatta fuggendo, massime quando il nemico havrà il pugnale.

p. 135 §2 In tutte le ferite si farà prima la lunghezza del corpo; & poi si soggiungerà col moto de i piedi; & si procurerà, che il punto sia veloce per arrivarsi alla ferita.

p. 135 §3 Ogni volta, che il pugnale pararà le nostre ferite, rifaremo le percosse; le quali ci aiuterà à fare la battuta del pugnale; cioè; noi non violentaremo il nostro braccio dal suo naturale.

p. 135 §4 Ogni volta, che accommetteremo con la nostra postura, se il nemico si ritirasse (ilche intraviene per lo più) noi abbassando l'armi di nuovo ricominciaremo la zuffa col nostro medesimo modo: & se pure ei parerà di stare in guardia, anderemo restringendo il nemico, quanto più potremo, mettendolo in necessita.

p. 136 §1

AVERTIMENTI DI SPADA SOLA.

p. 136 §2 OGNI volta, che s'habbia la spada impegnata, ella potrà liberarsi ritirando il piede manco adietro: & se il contrario ferisse, in quel tempo medesimo si ferirà.

p. 136 §3 Se tirandosi qualche ferita, si venisse à meza spada, liberaremo la nostra; se la teniremo di dentro, col tirarla per di sopra, piantando la ferita: & con la mano manca prenderemo la spada contraria nelli suoi fornimenti. Ma se l'havremo per di fuori, la caveremo per di sotto, cacciandosi dentro nell'angolo constituito dalle due spade col nostro braccio manco traversato alla difesa; & feriremo di punta col pugno della spada all'in sù.

p. 137 §1 # TRATTATO DI SPADA, ET CAPPA.

p. 137 §2 SEMPRE, che si serviremo della cappa per difesa, consideraremo che quella hà differenza dal pugnale: perche l'istessa può esser tagliata, & forata; ilche non aviene al pugnale: & perciò non pararemo mai con la cappa, nel modo istesso, come si faria col pugnale: & si come le coltellate col pugnale non si parano, cosi con la cappa non si potrà resistere à quelli: & poiche si cerca di tenere il pugnale libero dall'offese; cosi la cappa nõ[n] si dovrà tener sottoposta al nemico.

p. 137 §3 Tutte le medesime ragioni, & tutti i modi, c'habbiamo detto di sopra di spada sola; & di spada accompagnata dal pugnale, ci servono con la spada, & cappa: & di questo non si cava altr'utile, che d'impedire il nemico; affine, ch'egli non

p. 138 §1 [hab]bia cosi facile la ferita; & di coprire le parti da basso.

p. 138 §2 Et si come con la mano chiudiamo all'hora il ritorno alla spada nemica, quando uscendo coi punti feriamo; cosi con la cappa operando meglio otteniamo il nostro fine.

p. 138 §3 Et perche con la cappa privandosi di questa l'huomo, in due modi si puo al contrario nuocere, & impedirlo: l'uno è il getargliela nel viso: l'altro il buttar gliela sopra la spada nel suo più debole terzo; modo più facile del primo; perciò essa si piglierà con la mano piana, & col dito pollice solo; & tutta con un semicircolo per di fuori si raccoglierà nel pugno: accioche unita possa l'huomo à suo piacere, & servirsene; & anchora privarsene per venire alle prese.

p. 138 §4 Di quì può comprendersi, che s'havrà da tenere il me desimo ordine in cacciar mano alla spada, & in volgersi la cappa al pugno: come si fà nel modo di spada, & pugnale: & questo s'havrà da fare senza dar intentione al nemico della sua volontà: ilche sarà col procedere con la spada bassa, & con la cappa fuori del corpo, accommettendo con cautela; & distinguendo quello, che farà il nemico; ilquale, ò in stato, ò in moto si dovrà truovare.

p. 138 §5 Se si truoverà in stato il contrario, forzosamente sarà in uno de gl'atti, c'habbiamo già detto, però

P. 139 §1 [con]tra quelli useremo una di quelle offese, che di sopra habbiamo dichiarato.

P. 139 §2 Et se il medesimo contrario terrà la spada avanti, contra essa combatteremo co'modi già detti: ma egli tenendo la cappa avanti, sempre di moto nostro naturale à man diritta; con tagli & punte feriremo tuttavia quel pugno volto dalla cappa, discostandosi dalla spada del nemico.

P. 139 §3 Se poi peraventura noi truovassimo il nostro contrario in moto, noi opereremo contra lui, conforme alla proportione della ferita, che egli operasse: & essendo ò di taglio, ò di punta, l'istesse offese, & le difese, faremo, come se non havessimo la cappa: eccetto quando vorremo parare: che questo sempre non con la spada sola; ma con la spada accompagnata dalla cappa faremo, & di secondo tempo feriremo alle parti contrarie diritte.

P. 139 §4 Caso poi, che noi volessimo aspettare il nemico, avertiremo con qual piede egli si truovera avanti: & noi si presentaremo, con l'altro differente all'apposito: come, s'egli si truoverà col diritto avanti, noi col manco si metteremo all'incontro; s'egli col manco, & noi col diritto il simil faremo.

P. 139 §5 Avertendo, che quando il contrario si truoverà col diritto avanti, & che noi col manco si opporremo sarà all'hora con intentione di lasciar passare

p. 140 §1 tutte le ferite; & di ferire nella declinatione, con l'accompagnamento di cappa.

p. 140 §2 Et stando il contrario col manco avanti; & noi col diritto, subito opponendoci col manco poi pararemo, ferendo alle gambe overo al braccio diritto, overo di punta riversa.

p. 140 §3 In questo caso potremo parare i tagli con la cappa sola: perche con la cresciuta del nemico, & con la nostra pararemo nel primo terzo della spada, avanti che sia in declinatione in forza.

p. 140 §4 Si deve rimediare alla percossa nemica, con la nostra spada: accioche fuori di questa occasione si potiamo assicurare di tal ferita: & se ferissimo per ventura in quel tempo, per lo più si tirarebbe in parte, dove la cappa del nemico intricarebbe, & impedirebbe la nostra spada in modo, che non la riscoteressimo in tempo. Assicurando dunque noi stessi dalla ferita del nemico, lo feriremo di secondo tempo nelle sue parti diritte.

p. 140 §5 Per aspettare poi il contrario, si metteremo in questa guardia, che quì di sotto si vede; cioè; col piede manco avanti in mezo passo, col peso sopra la base diritta, & con la spada traversale; ia punta della quale superi il nostro capo; col pugno à livello della spalla manca, & con la mano

p. 141 §1 manca piana, involta nella cappa, & nel forte primo della spada messa.

p. 141 §2 In questa postura si terrà sempre la spada nemica dominata in dentro, & tirando egli qual si voglia forte di ferita (sia ò di taglio, ò di punta) solo col manco piede si crescerà in passo sforzato: & distendendo quella linea si ferirà, con l'aiutar la spada con la forza della mano manca.

p. 141 §3 Come dimostra questa figura.

p. 142 §1 [Illustration of sword and cape, a fencer thrusts with a left-foot-forward lunge]

p. 143 §1 CASO poi che si concedesse, che il contrario passasse à mano diritta, il rispondente servendosi del punto della prospettiva havrà da ferire d'imbroccata, lasciando la cappa; la qual chiuda il ritorno alla spada del nemico.

p. 143 §2 Se mentre, che noi si truoveremo in detta postura, il nemico non si risolverà à ferirci; ma mettendosi an ch'egli in guardia aspettarà, che noi siamo i primi à ferire; & esso si truoverà col pie diritto avanti; & con la spada in presentia, noi con la medesima postura lo stringeremo, ferendolo sempre con l'istessa ferita.

p. 143 §3 Ma se col manco piede posto avanti cercasse il contra rio d'adombrarci la nostra spada con la sua cappa, noi all'hora ritornando col manco piede nel centro, usciremo nella prima nostra guardia, tra le quattro c'habbiamo proposte; & à quel pugno di punta, overo di taglio feriremo: ò per di sopra, passando con la punta entreremo nel viso.

p. 144 §1 Buckler

TRATTATO DEL BROCCHIERO.

p. 144 §2 ESTATO da molti adoperato il brocchiero nella guisa, che si farebbe un pugnale; & con esso hā[n]no parato, accompagnandolo con la spada; ma perche difficilmē[n]te può l'huomo mostrar qualche facultà col mezo di qualche instrumento, se pra non si fà conoscere ciò che sia il medesimo instrumento; però diciamo, che co loro, i quali vogliono esprimere in una cosa alcuno effetto col mezo dell'instrumento à ciò appartenente, bisogna per ogni via, affine di non incorrere in errore, che essi conoschino la qualità, & pro prietà del medesimo instrumento, col quale s'intende di rappresentare quell'effetto. Havendosi dunque da trattare del brocchiero instrumento (come dissi) difensivo, cosi prenderemo il principio di trattarne.

p. 144 §3 Il brocchiero è corpo sferico, denso, & opaco: & affine ch'egli faccia bene l'offitio suo, sarà nella curva superficie difuori di diametro d'un piede: atteso che per il piu gli huomini sono larghi un piede, & mezo: & perche l'effetto del brocchiero è di

p. 145 §1 [resi]stere alla spada, accioche non possa ferire il nostro corpo; poiche con essa spada l'huomo può formare diverse, & infinite linee: le quali per similitudine potiamo assomigliare à i raggi del Sole, corpo sferico, & luminoso: perciò debbiamo sapere, che duè sono i corpi opaco; cioè; denso, & spesso: com è la terra; & diafano; cioè; trasparente: come sono l'aere, l'acqua, & il fuoco.

p. 145 §2 Onde i prospettivi, si come si legge nella vigesima quarta propositione della prima parte della prospettiva commune; vogliono, ch'un corpo sferico, denso, & opaco, & conseguentemente ombroso, in tre modi possa esser considerato, rispetto ad un'altro corpo sferico, luminoso, & risplendente.

p. 145 §3 Nel primo modo si considerarà, quando il corpo ombroso sarà maggiore del corpo luminoso: che all'hora vogliono, & pruovano, che il detto corpo ombroso verrà à causare, & mandare un'ombra, la quale quanto più lontano si distenderà, tanto più sempre s'allargherà, come se ne vede l'essempio nella seguente figura; nella quale il corpo luminoso sarà inteso per l'A, l'ombroso per il B, & l'ombra per il C.

p. 145 §4 [Illustration of the shadow made by an object larger than the source.]

p. 146 §1 Nel secondo s'havrà da considerare, che il corpo sferico ombroso sia della medesima grandezza, ch'è il luminoso: & all'hora l'ombroso manderà sempre l'ombra di ugual larghezza al suo diametro: estendasi quanto si voglia in lungo l'ombra: come potiamo manifestamente vedere nella figura, che segue.

p. 146 §2 [Illustration of the shadow made by an object the same size as the source.]

p. 146 §3 Ultimamente nel terzo modo si può considerare, che il corpo sferico ombroso sia minore del luminoso: & che cosi l'ombroso sia per causare, & mandare un'ombra, la quale, quanto più s'estenderà in lungo, tanto più si anderà restringendo fino à tanto, che ultimamente ella finisca in acutezza, & termini in un punto, dal qual risulti una figura piramidale: la punta estrema della quale si chiama Cono. Questa esperienza ci mostra la seguente figura.

p. 146 §4 [Illustration of the shadow made by an object smaller than the source.]

p. 146 §5 Di tutti questi modi il primo è quello, del quale noi si serviremo col Brocchiero: perche essendo egli

p. 147 §1 [cor]po sferico, opaco, maggiore del corpo sferico luminoso, ò maggiore del pugno della mano; dal quale, à guisa de raggi del Sole, sorgono le proportioni delle linee, le quali forma l'huomo con la spada, opporremo quello in modo, che manderà un'ombra, dalla quale noi saremo coperti, & difesi.

p. 147 §2 DISTINGUEREMO nondimeno in questa occasione, che il contrario overò havrà la spada in atto nella linea retta; overo in potenza; cioè; in linea obliqua: quã[n]do l'havrà in linea retta, questa consideraremo, come un raggio solo del Sole: & à quello nel punto istesso opporremo il Brocchiero: ma quando l'havrà obliqua al nostro corpo, opporremo il Brocchiero al corpo del pugno, come ad origine del raggio, overo della linea: che cosi saremo coperti dall'ombra del corpo opaco del Brocchiero.

p. 147 §3 Et si come molte linee rette, le quali vengono tutte da un centro, quanto più si allontanano da esso cẽ[n]tro, tanto più si dilatano fra se medesime: cosi opponendo noi il Brocchiero alla spada, ò quella in esso s'intopperà, ò quello segarà: ond'ella più si allontanarà dal suo centro, & dal nostro corpo.

p. 147 §4 Si terrà dunque in mano col deto pollice messo per

p. 148 §1 appoggio: accioche quella parte non ceda in dentro: ch'altramente facendosi il nostro corpo verrebbe discoperto, & offeso: & accioche sempre operi, come corpo sfecrio[1], lo terremo col braccio disteso: onde venga causata per l'allontananza d'esso dal nostro corpo l'ombra maggiore.

p. 148 §2 Di modo, che con questo, si come con tutte l'al tre sorti d'arme difensive, operaremo accommettendo, senza dare intentione al nemico della nostra volontà: distinguendo, che il nemico, overo sarà in moto, overo sarà in stato.

p. 148 §3 Se sarà in moto, ò tirerà di taglio, ò di punta: alle quali percosse di taglio, faremo l'istesse parate, & ferite; come habbiamo detto; ma parandole col Brocchiero, procederemo con l'arme accompagnate; & seguitaremo il centro del pugno, & col manco piede contrapasseremo.

p. 148 §4 Lascio di dire, che potremo ferire avanti tempo, & nel tempo, & dopo il tempo, con gl'istessi modi: come habbiamo detto nel trattato di spada, & pugnale: ma avertiremo di mettere sempre il Brocchiero appresso il pugno nostro della spada, distendendo bene il braccio, & il corpo; & coprendoci sotto à questo medesimo.

p. 148 §5 Se il contrario tirasse di punta, potrassi fare l'istesse ferite, & parate, che gia si sono dette: ma perche alle cose, mentre che sono in moto, facilmente

p. 149 §1 [s'ag]giunge nuovo moto, però si potrà far col Brocchiero l'istessa difesa, come se fosse un pugnale.

p. 149 §2 Quando poi il nemico si truova in stato, ò havrà la spada avanti in presentia, ò adietro: se l'hautà avan ti in presentia, ò la terra in linea retta, ò in obliqua.

p. 149 §3 A tutte le proportioni di linee, che saranno alte, noi gli opporremo nel punto suo estremo il Brocchiero; & feriremo per di sotto, con la nostra maggior lnnghezza,[2] & in una istessa battuta.

1 sic, mistyped *sferico*.

2 Sic, mistyped *lunghezza*. Imagine how many mistakes this book would have, without spell-check?

p. 149 §4 A tutte se linee rette in presentia, le quali si truovano al diritto del centro del nostro corpo(come quando sono in guardia terza, & quarta) si opporrà il Brocchiero nel suo estremo; & per di sopra metterassi la nostra spada nella ferita in linea retta.

p. 149 §5 Se il contrario terrà la spada in presentia obliqua al nostro corpo, si metterà il Brocchiero opposto al pugno; & con la medesima battuta si ferirà per di sopra al luoco piú vicino.

p. 149 §6 Caso, che il contrario tenesse la spada angolare, si farà l'istessa offesa, come se havessimo il pugnale, aiutando la spada del nemico alla sua inclinatione.

p. 149 §7 Mentre poiche truovaremo il contrario con la spada adietro, cercaremo il corpo nemico nel modo detto; & terremo il Brocchiero fuori del nostro corpo per battere, come se fosse un pugnale, necessitando la spada contraria à passar per quello.

p. 150 §1 Tuttavolta, che col manco piede avanti si fermasse il nostro contrario; & c'havesse il Brocchiero avanti posto, di moto nostro diritto naturale, arrivati in distantia, & col corpo in scurzo, cercaremo il nemico, allontanandoci dalla sua spada, & co'circoli andaremo cõ[n] la spada intorno al Brocchiero; e poi in arrivando al piano della nostra spalla, cacciaremo le punte[ ]per di sopra; & se quello alzarà il Brocchiero, per di sotto.

p. 150 §2 Potrassi anchora mettere la punta della nostra spada nel centro del Brocchiero del nemico, il quale si può tener certo, che vorrà abbassarlo, affine di veder detta spada: & in quel niedesimo tempo si ferirà, caminandosi con quella punta nel viso.

p. 150 §3 Per maggior intelligenza hò messo quì questa figura accioche si veda l'effetto che fanno le linee quando segano il brocchiero.

p. 150 §4 [Illustration of a hand holding lines which pass around, or are stopped by, a disc]

p. 151 §1

TRATTATO CONTRA UN MANCINO.

p. 151 §2 TUTTE le ragioni dette di spada sola; di spada & pugnale; di cappa; & di Brocchiero, servono cosi al Mancino contra al Diritto; come al Diritto contra al Mancino: &

Lefties

dell'arme da difesa non potiamo servirsi in altro, che in parare le botte, che di lontano sono tirate; & per chiudere il camino: perche venendo alla pugna il Diritto contra al Mancino tutti due hanno la spada da una parte, & il corpo posto in modo, che tutte le punte feriscono angolarmente: & perciò con l'arme difensive non si possono parare, onde converrà, che con la spada si parino: atteso che il debole della spada del Mancino passarà per il forte della spada del Diritto.

p. 151 §3 Terrà dunque la spada il Diritto in stato, che domini la spada del Mancino in dentro; & col corpo in profilo, & col filo buono si caccierà le punte al più prossimo luoco da ferire.

p. 151 §4 Il Diritto mai non ferirà per di dentro al Mancino; salvo se egli non fosse in moto, che in tal tempo si potrà ferire.

p. 152 §1 Havrà da distinguere il Diritto, se il Mancino terrà la spada avanti, ò adietro; & operarà contra à quella col modo detto; cioè; dominando la spada, quando ella si truoverà avanti, dando il moto al Mancino per ferire nel tempo; & quando si truoverà adietro, combatterà col corpo pronto à far l'istessa ferita, chiudendo il camino alla spada del nemico con l'arme difensive.

p. 152 §2 A tutti i tagli il diritto lascierà passarli; & ferirà doppo il tempo di fendente; overo gli aiuterà alla declinatione, & ferirà di riverso: si potrà anchora parare coperto, & farà tutte l'istesse ferite, che noi habbiamo detto, che si fanno avanti il tempo.

p. 153 §1 ABBATTIMENTO DI BARRIERA.

p. 153 §2 IN tutti gli essercitii Cavagliereschi, che si fanno per giuoco, il Cavagliero deve mostrare l'agilità, destrezza, & forza del corpo. Et sappiasi, che per due rispetti questa
Barrier sorte di spettacoli si suol fare: l'uno per mostrare con questo apparato à gli assistenti la similitudine d'un fatto d'armi: l'altro, affine, che ciascuno di quei cõ[m]battenti possa mostrar il valor della sua persona. Onde tra tutti i giuochi d'armi, quello della Barriera è il più nobile; ilquale non è altro, che rappresentamento di due battaglie: per le quali si viene prima all'urto della picca; poi si caccia mano alla spada; & con quella si procaccia d'ottener la vittoria; ch'è il suo fine.

p. 153 §3 Avanti, che si faccia la folla si combatte da corpo à corpo: & questo si fa, accioche ogn'uno in particolare non solo mostri la sua forza; ma ancho venga in cognitione lo spettatore della battaglia generale.

p. 153 §4 Imiterà dunque il cavagliero in tutti gli essercitii cavagliereschi il vero, come se fosse nel fatto istesso co'ferri ammollati; fuggendo nondimeno tutti gli

p. 154 §1 estremi; ne essequendo le cose totalmente da dovero, ne totalmente per giuoco; & procurando di far tutti i suoi moti naturali: cosi di tutto il corpo, come delle parti d'esso: & questo sempre senza affettatione alcuna,

p. 154 §2 *In questo combattimento concorrono due sorti d'arme, l[a]. picca, & lo stocco, overo la spada: delle quali hora noi in particolare tratteremo.*

p. 154 §3 LA picca hà tre proportioni, cioè; il piombo, ch'è; il perpendicolare, la medionale, & l'orizonte; il piombo è, quando stà perpendicularmente: la medionale, quando dal perpendicolare si parte; & và nella declinatione: l'orizonte, quando si truova per linea retta dall'occhio riguardante ad alcun punto apparente, che sia nel piano dell'orizonte.

p. 154 §4 Sei cose s'operano con la picca: la prima è, che con essa si passeggia alquanto il Campo. La seconda è, che con essa s'appresenta alla battaglia. La terza è, che con essa si passeggia per andare alla ferita. La quarta è, che con essa si ferisce. La quinta è, che essa si getta in terra, per poter mettere mano allo stocco.

p. 155 §1 La sesta, & ultima è, quando partendosi dal Campo si camina con essa inarborata.

p. 155 §2 Il passeggio, che l'huomo con la picca fà, quando comparisce nel campo, hà da esser tale, che il Combattente mostri scioltezza nelle gambe, inerbandole; & fortezza nella vita, sprezzando il peso dell'armi. Dovrassi avertire di fare i passi ne grandi, ne piccioli; ma naturali, & senza affettatione alcuna. Tengasi posta la picca su la spalla diritta, col calce d'essa sotto l'orizonte; & in modo, che tirandosi una linea dal ginocchio si formi con essa angolo acuto: & tengasi il braccio diritto, non nel piano della spalla; ma un poco pendente per fuggir l'affettatione.

p. 155 §3 Et ogni volta, che il Cavagliero si levarà la picca di spalla per inarborarla, ciò farà co debiti modi, non facendo moto alcuno, che non sia fatto à tempo; & concertatamente. Questo ottimamente riuscirà; se nel levar la picca; si alzarà il piede diritto; & nell'abbassare la medesima si tirarà adietro il diritto, con la visuale verso la punta; & parimente nel levar la picca, per rimetterla su la spalla, si levarà il piede, con battere sempre l'istessa battuta.

p. 155 §4 Il presentarsi alla battaglia si farà col corpo disposto, & diritto; & perche quiui in stato si suole aspettare, che il Mantenitore sia in ordine; in questo tempo

p. 156 §1 dunque si farà qualche moto; hora con la testa; hora con un piede; hora col braccio manco; & hora con tutta la vita: & questo per mostrarsi vivace; & accioche non si paia una statua: la picca: oltra di ciò si terrà à piombo col calce appresso il piede diritto; & col braccio diritto disteso in alto: & come dimostra questa figura, nella quale vi è la proportione del corpo, che partendosi dalla quiete si fa nel passaggio alla nostra manca mano, & l'altra che nel voltarsi al passeggio à mano diritta si fa, qual mostra il corpo in prospettiva.

⁂

p. 157 §1 [Illustration of three figures holding pikes. From left to right:]

Postura. Passeggio à man diritta. Volta.

p. 158 §1 IN partirsi dallo stato per entrar nel passeggio, che è la terza operatione, che si fa con la picca, q̃[ue]sto si farà traversalmente, cercando di guadagnar terreno, & d'approssimarsi al nemico. Et poiche il corpo nostro hà tre ꝓ[ro]portioni; cioè; la prospettiva, lo scurzo, il profilo; si operarà in modo, che la picca col corpo formino sempre l'istessa veduta: & perche non si può di prospettiva entrare in profilo, senza passare per il mezo; ch'è lo scurzo: però sempre, per fuggir l'errore de gli estremi, dall'una nell'altra proportione s'entrera con quest'ordine da noi mostrato nelle figure precedenti, si cercarà nondimeno di tenere per il più il corpo in scurzo, affine di formar più bella vista: & la picca si terrà nella linea medionale, come hanno dimostrato le duè figure, le quali hã[n] no le vedute, che fa il nostro corpo; cioè; il scurzo, & la prospettiva.

p. 158 §2 La quarta operatione, che si fà con la picca, è la ferita, laquale conforme à i capitoli stabiliti trai combattitori, sarà da farsi in un particolar luoco del corpo, se bene per lo più si constituirà dalla Goletta in sù; & la picca, per tenersi nella sua forza, bisognerà, che si truovi in linea retta; e nella sua terza proportione orizontale: come si vede nella seguente dimostratione, nella quale si mostra prima il profilo, che fa

p. 159 §1 il corpo con la picca quando passeggia à mano diritta; & poi il secondo modo quando rivolgendosi il corpo per entrar nel scurzo, ilquale fa l'huomo ò per abbassar la picca alla ferita, ò pur per passeggiare, & la terza figura è la ferita della quale qui si parla.

⁂

[Illustration of three figures holding pikes. From left to right:]

Passaggio à man manca volta. Ferrita.

p. 161 §1 ET perche la vera operatione di picca si fà col mezo del peso del corpo, dovrà il Cavagliero con una forza, & non più, far detta ferita: la qual forza sarà di tre passi andanti: che tã[n]to importa la lunghezza dell'huomo, il quale eccedẽ[n]do la detta quantità ricominciarà un'altra forza.

p. 161 §2 Del modo poi di rompere la picca eleggerassi il migliore; potendosi in quattro modi rompere. Nel primo, si può mettere il casce d'essa nel fianco manco, per avanzare lunghezza: si può ancho rompere cacciando il pie diritto al manco: ma oltre, che questo fà scomparuto vedere, difficilmente poi si può avertire il punto, per truovarsi la picca fuori della nostra visuale: ne una picca facendosi da dovero si terrebbe in tale stato: ne cosi operandosi, il corpo è in proportione d'usar bene della sua forza

p. 161 §3 Ma quando si fosse mantenitore si potrà meglio, & con più ragione variare ferendo in questi altri modi, che seguono.

p. 161 §4 Nel secondo si può col moto delle braccia, discostandole fuori del corpo, ferire, rompendo la picca nel corpo: atteso che difficilmente si può accogliere in altra parte: ma tutte le percosse sono oblique, & di molto sconcerto: ne questa è la vera operatione del la picca.

p. 162 §1 Nel terzo (& questo hà più del perfetto) si può co tre passi della forza, passar dal diritto piede, & ferire con una mano sola.

p. 162 §2 Nel quarto, & ultimo, (il quale è quello, di che si serviremo, fatto che si havrà il passeggio col corpo in prospettiva) à poco à poco si calerà la picca, laquale sara ricevuta dalla mã[n] ca mano: & arrivatosi poi in distanza, cõ[n] essa nell'orizonte, s'avertirà, che il suo estremo sia per diritto al punto, nel quale si vorrà ferire: & ritirandosi il pie manco appresso al diritto, & aggiustando la ferita, si farà il moto de i tre passi, movendo il manco piede prima, & poi il diritto: & finalmente si farà finir sopra il manco la declinatione del peso, & il moto della forza dell'huomo: si terrà ancho il braccio diritto in angolo; & il pugno appresso il cinto, per operar la ferita in linea retta.

p. 162 §3 La quinta operatione, che si fà con la picca, è la ritirata, doppo che si sarà fatta la ferita: & questo per rifar di nuovo battaglia con un'altra picca; overo per cacciar mano allo stocco; la qual picca, rotta c'havrà il Cavagliero, la gettarà in terra. Et questo gettarla in terra si fà in tre modi: Nel primo in dietro, ò per fianco, ò per di sopra la spalla diritta: ma questo, oltre che cõ[n] la picca da dovero non si può fare, per lo più si dà in testa, & à padrini, & à i circonstanti.

p. 163 §1 Nel secondo si può buttare la picca per disprezzo contra il nemico: ma essendo in potere del medesimo di rimandare anchor esso il suo scalzo, à parer mio, non è da vilipendere il nemico; poiche incontinente si havrà da fare à colpi di stocco, ne quali consistendo il maggior valore, altri può ricordarsi dello sprezzo, & render pane per focaccia: perche dove s'attende à fatti; poco sono curati gli sprezzi.

p. 163 §2 Nel terzo, il quale noi bene useremo, sarà, che subito c'haveremo fatta la ferita, col manco piede ritirandoci, alzeremo il calce della picca, & col diritto cõ[n] nuovo passo ritirandoci, questo medesimo casce faremo sdrucciolare per la mano manca; & per di sotto il fianco nostro diritto la gettaremo in terra: ilche, oltre che fa bellissimo vedere, s'approssima poi più al verisimile: perche cosi, fatto che sia l'urto cõ[n] la picca, si viene alla folla, & se la lasciamo cadere in terra; & subito mettiamo mano alla spada.

p. 163 §3 La sesta operatione è, che partendosi il Cavagliero dal Campo, finita che sarà la battaglia, & ritruovandosi esso in stato con la picca inarborata nel pendiculare, si metterà in moto col manco piede; & poi col diritto appresso al medesimo, nell'istessa battuta metterà il calce della picca: & fatti circa à cinquc, ò sei passi, distesi co' debiti modi, & abbassando la picca se la metterà in spalla, & seguiterà il suo camino.

p. 164 §1 TRATTATO DELLO STOCCO.

p. 164 §2 IL Cavagliero nel bat tere di stocco haverà tre parti: cioè; scioltezza, forza, & prestezza: la scioltezza consistem in liberare bene i colpi: ilche è di grande giovamento: peroche non si lascia intricato lo stocco con quello del contrario: la forza con tutto che sia dono di natura: nondimeno perche la percossa dello stocco si fà col moto, bisogna star su l'aviso d'eleggere il migliore: che con questo mezo, si batterà con maggior forza: la prestezza, per esser nel fatto dell'armi dono pretiosissimo di natura, farà, che'l Cavagliero sia ben risoluto, in tutto quel modo, ch'egli vorrà operare in battere; & cio ancho farà con molta brevità di tempo.

p. 164 §3 Et perche à pieno s'è trattato della natura del moto, &, di quella de'tagli; & quali d'essi siano maggiori, conforme al loro descenso, non occorre, che quì più à lungo ci stendiamo: solo diremo, chel moto naturale aiutato dalla possanza movente, farà maggior effetto in un resistente; & che la vera percossa sarà, quando lo stocco col braccio,

p. 165 §1 formarà ferendo linea retta con la spalla.

p. 165 §2 Et in più modi si può battere: Nell'uno con mandiritti tondi; nell'altro con mandiritti traversali, & con diritti raddoppiati; overo si può battere à braccio aperto, & si porta con la spalla il peso, facendosi moto dal centro del corpo in su.

p. 165 §3 Col battere co'tagli tondi non si libera bene lo stocco: & la percossa non è la maggiore, che si faccia.

p. 165 §4 Quando noi colpiremo co'traversali, aiutati dal maggior moto, & peso del corpo, sarà il miglior modo, che si faccia: perche con questa via si percoterà in mezo alla tempia, & la mascella, ch'è il più facil luoco per stordire il contrario, & per aiutarlo à declinare.

p. 165 §5 Il battere col quinto cẽ[n]tro di ridoppio si fà per questa medesima causa, accioche, essendo dette percosse di poco tẽ[m]po, pini presto si stordisca il nemico: ma queste sono di poco peso; & si và à pericolo d'incorrere in qualche disordine.

p. 165 §6 Quando poi si porta col moto del corpo il braccio aperto; & che si batte nel mento, quest'atto è feminile; & è di gran tempo, di modo, che ne potrebbe cader lo stocco, oltre che fa brutto vedere.

p. 165 §7 Resta dunque, che il nostro Cavagliero sia il primo à ferire il nemico: perche egli è sempre bene, & avantaggio l'essere il primo feritore: & questo con un mandiritto traversale, facendo centrò. del piede

p. 166 §1 manco: accioche la circonferenza sia maggiore. Di più restarà ingannato il contrario della distanza; & resterà ferito, mentre che si truoverà in moto.

p. 166 §2 Quando il Cavagliero s'avedesse, che alcuno de combattenti, fosse sconcertato nel menare i suoi colpi, egli potrebbe presentarsi in stato per ferire di riverso; overo essere il primo à ferire di taglio diritto, facendo questo nondimeno debilmente; & poi quando il contrario ferisse di diritto col riverso gagliardo, durerà aiutarlo alla declinatione, dove penderà il suo peso.

p. 166 §3 Et è da notare, che colpendosi il contrario, il quale si vegga sconcertato dalla percossa, si farà l'istessa botta: com'ancho se col diritto lo vedremo scõ[n]certare noi raddoppiaremo il diritto. & il medesimo si farà del riverso.

p. 166 §4 Caso, che gli stocchi nel menar le mani s'intricassero, bisognerà avertire di non tirare verso se; ma verso terra: nel qual tempo bisognerà mettere il ginocchio manco sotto la barriera; & con ambedue le mani aiutarsi.

p. 166 §5 Et se, come suole avenire, il contrario cedesse, & col corpo venisse sopra la barriera, con la mano manca si farà cadere verso se; se quella sopra l'elmo si metterà aiutandolo al centro.

p. 166 §6 Si potrà in due modi far cadere lo stocco al contrario:

p. 167 §1 Nel primo, quando egli ferisse di taglio, percotendolo nel braccio pur di taglio: Nel secondo, quando egli volesse formare il riverso, se noi metteremo il nostro stocco con la punta sopra la sua spalla diritta, & con quello formaremo una linea.

p. 167 §2 MA per venire più alla prattica, il Cavagliero dovrà sapere i capito li; & le constitutioni fatte da'Mã[n]tenitori: accioche possa operare conforme ad essi: & cosi con l'ordine de paggi, de padrini, & de tamburri entrerà nel campo: del qual ordine, per esser noto, non dirò altra cosa.

p. 167 §3 Ma in passeggiando esso arditamente il campo, giuntò, che sarà al conspetto delle Dame, inarborando la picca, à quelle farà riverenza, col modo già detto. Poi seguendo farà il medesimo avanti à i Giudici, se ben questo tutto con minore humiltà.

p. 167 §4 Giunto poi, che si sarà al luoco deputato, per fare il giuoco, il Cavagliero inarborando la picca con i debiti modi, in stato si presentarà alla battaglia; & si mostrerà vivace nell'armi. Il quale nel levare il piede diritto, levarà insieme la picca in alto con la mano diritta; & con la manca la piglierà lungi dalla diritta in honesta distantia; & ritirando il pie diritto adietro farà riverenza alle Dame. In ultimo

p. 168 §1 col finir la battuta il medesimo piede, la picca finirà la sua declinatione traversale.

p. 168 §2 Poi in rihavere il piede diritto il Cavagliero, rihavera la picca, passã[n]do col piede, & cõ[n] la picca in prospettiva, & cõ[n] la battuta del diritto; col corpo in scurzo, & la picca in detta proportione si pigliarà solo la picca con la mano diritta; & alzandosi il piede mã[n]co, s'alzerà la picca in alto; & in questa facendosi atto di riverenza col manco, si basserà la picca; & pur col manco entrando nel passeggio, & col corpo in scurzo, s'alzarà la picca in alto; & si terrà nella medionale, sempre battendosi l'istesse battute col piede, & con la mano.

p. 168 §3 Poi, quando si sarà arrivato al capo del passeggio, batterà il Cavagliero una battuta, riguardando il contrario, ciò ch'egli faccia: & vedendo, ch'anchora esso non entri alla battaglia, si rimetterà nel medesimo passeggio con far un moto di vita indicativo, che la volontà di combattere era pronta. Ma vedendosi in fine venire il mantenitore alla battaglia, col modo già detto esso anderà à ferirlo: & fatta la ferita si ritirerà, ripigliando l'altra picca. Cosi di mano in mano operarà; & poi c'havrà rotto l'ultima picca, cacciera subito mano allo stocco.

p. 168 §4 Et perche il tutto proceda concertatamente, si deve sapere, che in vece della ritirata, che si fà in alzar il piede manco, & poi nel ritirarlo, s'alza la picca; & poi

p. 169 §1 ritirando il diritto, si getta la medesima per di sotto al fianco diritto. In quest atto poi, che sifa nell'alzare il manco, s'alza la picca, & nella battuta del piede si getta la medesima per di sotto al fianco manco: & per di sotto alla spada aiutata dalla mano manca, & poi col diritto adietro si caccia mano nel medesimo tempo allo stocco; & seguitando la ritirata s'opera nella maniera, che qui segue.

p. 169 §2 Si farà riverenza col pie diritto, & col corpo in scurzo volto alle Dame, tenendo lo stocco traversato per di soprà con la mano manca. Poi passando col pie diritto in quella battuta s'alzerà lo stocco in alto, voltandosi à i Giudici; & col pie manco facendosi atto di riverenza. Poi con prestezza ripassandosi col pie manco si ripiglierà lo stocco nel suo estremo con la mano manca per di sopra: & riguardandosi la barriera, si appresentarà il Cavagliero distante due passi in stato di ferire col manco piede avanti, & mentre ch'egli arriverà ritenendo il fiato con tutta la sua forza il contrario ferira di taglio traversale: & replicando il riverso finirà i suoi colpi: i quali, forniti che sarã[n]no, esso ritirerà il piede diritto adietro un passo; & con lo stocco posto in alto in atto di ferire, farà un moto di testa con bravura.

p. 169 §3 Si potrebbe forsi dubitare, se con lo stocco si debba fare la riverenza: perche rotto, che s'è la picca, si propone senz'altro la battaglia con lo stocco. Ond'io

p. 170 §1 rispondo, che, perche la picca, & lo stocco sono armi differenti; & perche per ciascheduna di loro si dà premio; & perche il far riverenza dà tempo à gli spettatori, di prepararsi alla vista del rimanente: però questo sarà da farsi, se bene alla battaglia ciò sia per importar poco.

p. 170 §2 Si ritirerà ben poi sempre il Cavagliero col corpo in prospettiva, & co'passi detti di sopra, tuttavia vibrã[n]do lo stocco; il quale, giunto ch'egli sarà al padrino, subito rinfodrerà, per cedere il luoco à gli altri combattenti.

p. 170 §3

TRATTATO DEL LA FOLLA.

p. 170 §4 SI rompe, secondo che si concerta una, overo tre picche nel modo detto disopra: fatto questo si viene alla zuffa di stocco; & volendosi contrapassare, bisogna, che'l primo si metta à man diritta: accioche si possa con le parti manche spingere i suoi compagni. Et sempre s'avertirà di ferire di man diritto; atteso che ferendosi di riverso non si potrebbe riuscir bene per

p. 171 §1 [l'impe]dimento delli compagni: & tutte le botte cosi de nemici, come de gli amici accoglierebbono lopra quel braccio. Et questo è quanto m'occorredi dire sopra il combattimento, che si fà nella Barriera.

p. 171 §2 Mi resta nondimeno d'avertir gli Padrini, che, quando danno la picca in mano al suo Combattente, debbano prima pigliar la loro mano diritta con la manca loro: & cosi porger loro la medesima picca, avertendo essi di metterglièla in mano, con il calce commodo: accioche il Combattente nõ[n] habbia causa di mettere tutte due le mani per accommodarsela ben nella sua mano diritta: il che è brutta cosa da vedersi.

p. 171 §3 I Tamburini, quando arriveranno al Campo, dovranno battere la chiamata: quando entraranno, l'ordinanza: quandos'incominciarà il passeggio, batteranno pur chiamata: quando si abbasserà la picca alla ferita, battaglia: quando poi si sarà rotta la picca, la ritirata: & cosi farassi di mano in mano, secondo il bisogno. Questo medesimo ordine s'havrà da servare, quando li combatterà con lo stocco.

p. 171 §4 Nõ[n] hò voluto qui trattare di certe vane levatè, le quali talhora per galla si possono fare: percioche l'inventione di questo appartiene à gli Mantenitori: perche, quando si procede con le ragioni, & regole, ch'io hò dette di sopra, da se le potranno fare, &

p. 172 §1 comporre à suo modo i Cavaglieri: i quali nondimeno studieranno di tenersi ſempre più, che ſia possibile, dalla parte del verisimile.

p. 172 §2

TRATTATO D'UN COMBATTENTE A PIEDI CONTRA UNO A CAVALLO.

p. 172 §3 Horse VOLENDO l'huomo à piedi contrastare contro uno, che combatta à cavallo, conviene, ch'egli operi conforme alla natura del cavallo, il quale essendo formato di corpo lungo sopra quattro basi, hà per moto naturale il caminare per il diritto: & caminando con ogn'altra sorte di moto, quel moto li sarà violento: attesoche volendo esso muovere le parti d'avanti ò alla diritta, overo alla manca mano farà centro delle basi di dietro: & per il contrario, quando muoverà le basi di dietro, farà centro delle basi davanti.

p. 172 §4 Essendo dunque naturale di tutti gli animali il fuggir le cose, ch'apportano danno al ben'essere loro (percioche sono mossi dalla concupiscevole potenza, à questo

p. 173 §1 fine conceduta loro dalla natura) di qui nasce, che il Cavallo percosso in qualche parte del corpo, cede da quell'istessa banda per allontanarsi dall'offesa: & se bene sarà offesa ne fiã[n]chi da gli sproni, quando nondimeno egli sarà percosso in un'altra parte del corpo, che de'fianchi sia più nobile, con maggior offesa di quella, che fanno gli sproni, cederà da quella parte, nella quale riceverà maggior offesa.

p. 173 §2 Il nemico dunque posto à cavallo cercherà di urtare il contrario, mettendo il cavallo in fuga: overo cercherà di levare al medesimo la mano diritta, per poterlo ferire con la sua spada. Ma colui, che combatterà à piedi, cercherà di fuggir quella parte, col cercar sempre il nemico dalla mano manca di lui; & dalla sua propria diritta, affine d'allontanarsi dalla ferita nemica, & approssimarsi al contrario: come dimostra questa figura.

p. 174 §1 [Illustration of a man on foot with a sword and cloak facing off against a horseman]

p. 175 §1 PARIMENTE, perche il cavallo non potrà mai se non per il dirittò urtarci di moto suo naturale, avertiremo di presentarci sempre al diritto del cavallo: che cosi impediremo la vista al nemico, il qual si truoverà à cavallo: & fermi in mezo passo, & uniti in forza, & posti cõ[n] la spada distesa, fuori del nostro corpo, & co la cappa imbracciata nel pugno in modo, che bisognando se ne potiamo privare, subito, che il cavallo arriverà in distantia, tiraremo un taglio nel muso d'esso cavallo; overo una pũ[n]ta in quel medesimo tẽ[m]po, passã[n]do dal piede diritto, seguito dal manco alle nostre parti diritte; & ferendo d'un riverso traversale nelle redini. Poi mettẽ[n]dosi nel cẽ[n]tro della circonferenza, che forma il cavallo in stato, quã[n]do cõ[n] un circolo si giri il suo corpo, qui staremo sẽ[n]za più abbã[n]donar il nemico, cõ[n] ferir di continuo ò il cavallo, ò l'huomo.

p. 175 §2 Potrassi anchora in quel tempo, che il cavallo arriverà in distanza, spaventarlo cõ[n] la cappa; overo tirarglie la sopra la testa, col far l'istesse passate, & ferite.

p. 175 §3 S'avertirà ancho di dar della mano manca nella briglia; overo con la medesima mano dargli lieva sotto il suo piede; & cosi scavalcarlo.

p. 175 §4 Et caso, che il cavallo contrapassasse, se gli doverà all'hora tirare un taglio à i garetti.

p. 176 §1

TRATTATO D'UNO A CAVALLO,

p. 176 §2 *che con la spada combatte contra un'altro pure à Cavallo.*

p. 176 §3 LA medesima ragione, che noi usiamo in guadagnare la linea retta, mientre che siamo à piedi, la medesima useremo, mentre, che siamo à cavallo. Et questo piú facilmente: perche il moto del cavallo sara con più tempo fatto, che non era il moto nostro, quando eravamo à piedi; & il cavallo non serve all'huomo in altro, che per la commodità delle gambe del Cavagliero.

p. 176 §4 Et perche il cavallo è di corpo lungo, forma com'una linea: laquale, quando si porta di moto traversale con uno de'suoi estremi stà ferma; con l'altro forma un circolo: cosi il cavallo di moto violento, facendo centro delle basi di dietro, col restante del corpo forma un circolo.

p. 176 §5 Et si come à piedi guadagnandosi il diametro del circolo, si guadagna la linea retta; & si necessita il nemico à passare per essa: cosi stando à cavallo,

p. 177 §1 [sem]pre guadagnaremo il diametro del circolo: & accioche diamo qualche cognitione di questo, cosi diremo.

p. 177 §2 Si doverà sempre andar contra il nemico per il diritto; cioè; à testa per testa: & se bene il contrario andasse traccheggiando, questo faciliterà molto più il nostro fine: perche cosi con maggiore agevolezza si manterremo nel centro col nostro cavallo, formando circonferenza molto minore di quella, che farà il nemico col suo: & avertiremo ancho di mã[n]tenersi sempre nella linea retta. Ma presupposto, che il nemico venga all'urto, arrivato ch'egli sarà nella distantia, da qual si voglia parte, che il cavallo volgerà la testa, noi entraremo per il diritto, e terremo nondimeno la spada col pugno avanti l'arcione, & traversata in modo, che la punta guardi l'orecchia manca del cavallo: che in tal modo è in stato per parare, & ferire: non si facendo standosi à cavallo altre ferite, che queste due; cioè; una punta di sotto in sù al lungo del collo del cavallo; & il riverso per l'istesso camino: come dimostra questa figura, nella quale quella linea tirata mostra la postura; & la spada la ferita.

p. 178 §1 [Illustration of two men with swords on horseback at right angles]

p. 179 §1 ET caso, che il contrario voltasse, & pigliasse la carica, lo seguitaremo per retta linea cõ[n] la testa del nostro cavallo alla culatta del suo: & quando si voglia voltare si cacciaremo nel diametro: anchora che, quando havremo sotto buon cavallo, si potrà incodarlo, mettendoselo sotto la mano diritta: onde noi alla testa, overo al fianco del cavallo, potremo ferirlo.

p. 179 §2 Ritruovandosi il Cavagliero sopra un cavallo, ilquale non sia sufficiente all'urto, egli potrà risolversi di fuggir l'incontro, incontrando anchora esso il nemico, il qual venga alla sfilata all'incontro; & mettendo il suo cavallo in fuga, avertirà di mettere il braccio diritto disteso in linea retta, & in forza: che cosi facendo sempre truoverà il contrario, il qual havrà il braccio fuori di forza: & di qui facilmente potrà fargli cadere la spada, overo lo stocco.

p. 180 §1 *L'essercitio di correre le lancie, fu inventato cosi per dar spasso, & inanimire i Cavaglieri; come per renderei medesimi prattichi, & instrutti in dette operationi: accioche habbiano da essere esperti per quando si verrà al far da dovero. Onde per facilitar questo modo, fu inventato il correr all'Anello: se bene si costuma anchora di rō[m]pere le lancie nella Quintana; ò all'incontro dentro alla lizza; over'essercitarsi senza quella. Ma p[er]che tutte queste maniere d'essercitii si fanno ad un solo fine, il quale è di sapere ben portar la lancia, per servirsene à campo aperto, & alla guerra: & perche tra tutti questi giuochi quello del correr all'Anello è il più commune, et il più vasto, di quello dunque alquanto trattaremo.*

p. 180 §2 PRIMIERAMENTE sei sono le circonstanze, le quali concorrono al ben correr lancie. La prima è tener ben la lancia sopta la coscia. La seconda è levarla con attitudine da quella. La terza è l'arrestarla garbatamente. La quarta è l'abbassarla à tempo. La quinta è il ferire all'occasioné. La sesta è il riscuoterla come conviene.

p. 180 §3 Nella prima. In tre modi si può tenere la lancia sopra la coscia: avertendo prima il Cavagliero, che

p. 181 §1 quando egli sarà armato, dovrà tenerla fra la coscia, & l'arcione: atteso che l'arnese non concede, ch'ella sopra lui si debba posare. Nel primo modo dunque la lancia si tiene con la punta verso le sue parti destre all'Albanese: & questa domandasi lancia aperta. Nel secondo si tiene con la punta à mano manca: ma noi per fuggir gli estremi sa doveremo tenere nel mezo: accioche ella faccia col corpo l'istessa visuale. Nella medionale la terremo col braccio diritto in stato, che formi angolo col gomito; & questo, ne alto, ne basso; ma naturalmente in modo che potiamo sentirsi commodi in forza, & questa sarà la terza maniera.

p. 181 §2 La seconda circonstanza, qual è il levarsi la lancia dalla coscia, può in sei modi farsi. Nel primo ella si leva col pugno in presenza all in sù; & reggendola di polso a poco à poco s'arresta. Nel secondo ella si leva distendendo il braccio all in giù al lungo del fianco diritto. Nel terzo ella si leva allargando il pugno dal corpo col braccio disteso all'Albanese. Nel quarto ella si leva in alto, & subitos arresta alla Stradiota. Nel quinto ella si leva col braccio allungo del fianco diritto, ma un poco piegato, & cō[n] la mano volta in dentro. Nel sesto (& questo sarà il modo, col quale noi opraremo: percioche egli è il migliore) ella si leva dalla coscia, lasciandosi il pugno in quel piano, & di polso reggendola si mette

p. 182 §1 il punto d'essa al diritto dell'Anello; & si tiene il braccio piegato alquanto: che cosi ella viene sostenuta più facilmente; & è più propinqua ad arrestarsi con maggior facilità: & questo è secō[n]do l'uso della guerra.

p. 182 §2 La terza circonstanza, ch'è l'arrestarsi la lancia, si fà in quattro modi. Nel primo s'arresta nel cominciar la carriera. Nel secondo, quando è inviato il cavallo. Nel terzo, quando il Cavagliero s'avicina al suo aversario distante una lancia di lunghezza. Questi tre modi, oltre che fanno brutto vedere: & se bene quest ultimo si può pur fare, nondimeno quando si corre con la lancia incassata, per il moto del cavallo facilmente la lancia si dimena, dovendosi ella portar ferma, ne risulta bruttissimo effetto. Il modo d'arrestar si può fare con la mano volta all'in sù, quando si corre disarmato, & volta all'in giù: & questo modo è tenuto il migliore, per esser moto naturale, si come ogni altro è violento. Nes quarto s'arresta nell'istesso tempo, che la lancia arriva in linea retta nella ferita.

p. 182 §3 La quarta circonstanza, ch'è abbassar la lancia, in quattro modi si fà. Nel primo, subito, che s'è arrestato si abbassa. Nel secondo, quando l'huomo s'avicina al nemico s'abbassa un poco. Nel terzo s'abbassa, & si ferisse in un tempo medesimo; il primo non conviene farsi in nissuna maniera; il secondo è più

p. 183 §1 comportabile; il terzo fa bellissimo atto, quando si possa fare; ma è difficile. Il quarto, ch'è il migliore di tutti, si fa con l'abbassare à poco à poco la lancia, raccolto, ch'ella s'havrà in modo, che finisca la declinatione nella ferita.

p. 183 §2 La quinta circonstantia, che è la ferita, se vorremo accertare il punto, subito in partirsi, metteremo la lancia traversata in modo, che il suo estremo sia al diritto del punto, il quale si dovrà ferire: & che la nostra visuale, passando per difuori della lancia, scuopra il punto, overo l'Anello.

p. 183 §3 La sesta circonstanza, laqual è il recuperare la lancia, in due modi si fa. Nel primo ella s'alza in alto, & si rimette sopra la coscia. Nel secondo si getta il calce della lancia di dietro alla destra coscia, facendo che la punta d'essa guardi pure in dietro; e poi, quã[n] do è fermato il cavallo, si rimette sopra la coscia. Et io distinguerei questi due modi; che correndo all'Anello, fatta che fosse la ferita, si dovesse alzare il pugno in alto; & parato il cavallo, rimetterla nella coscia: perche la lancia integra si deve tenere in atto di ferire: ma se si rompessero le lancie, io loderei, che il tronco d'esse si mettesse di dietro alla coscia diritta.

p. 183 §4 Et tirata una linea retta, la quale venga dall'Anello, con la linea della carriera, dovranno formarsi due linee parallele: & la linea della lancia nell'Anello

p. 184 §1 ferendo formerà angolo acuto con la linea data: come mostra questa dimostratione quì di sotto: perciò si terrà la lancia sempre in tal proportione, che cosi sarà in atto di ferire col minor moto, che sia possibile: Metterassi dunque l'Anello distante dalla linea della carriera tre piedi di misura, & alla mano manca del Cavagliero che correrà. Di più si porrà alto da terra sei piedi: ma per l'ordinario ciascuno, che vorrà correre pruovera prima l'altezza dell'Anello: la quale dovrà esser tale, che il Cavagliero possa toccare l'Anello con l'estremità delle dita: che in tale stato il corritore si assuefarà meglio à correre la lancia alta: di modo, che venendo esso poi all'incontro ò nella lizza, ò à campo aperto colpirà nella testa facilmente, ch'è il più notabil luoco da ferirsi.

p. 185 §1 [Illustration of a man tilting at a ring]

p. 186 §1 LA lancia doverà esser lunga in tutto diece piedi, il calce suo infino alla impugnatura si farà lungo un piede, & un quarto: acciò ch'egli sia commodo. Ultimamẽ[n]te si farà il medesimo calce grosso in modo, che la sua circonferenza intorno all'impugnatura sia di sei oncie.

p. 186 §2 Et perche quest attione di correr lancie principalmente consiste in sapere il Cavagliero servirsi del cavallo: percioche da quello dipende quasi il tutto; anchora che questa sia professione di cavallerizzo; à bastanza nondimeno del proposito nostro qui ne parleremo.

p. 186 §3 Dovrà dunque il Cavagliero conoscere la natura del cavallo, sopra il quale egli correrà; cioè; se sarà per riuscir facile alla carriera, ò se difficile: quando sarà facile non occorrerà, ch'egli lo sproni; poiche quasi da se pigliarà la fuga dal pugno della briglia. Di que sta sorte i cavalli sono perfetti. Ma quando il cavallo sarà difficile, potrà il Cavagliero accostarli le gã[m]be in maniera, che con ogni poco moto esso lo pũ[n]ga: ma questo cosi cautamente farà, che i circonstã[n]ti non se n'accorghino. Et per dar la fuga al cavallo col pugno è da sapere, che à quei cavalli, che sottomettono il capo, si deve porgere avanti il pugno tanto, che giunga sopra il collo; per quelli poi, che

p. 187 §1 portano la testa alta, si deve abbassar la mano, che tiene la briglia. Ma bisogna bene avertire, che non sia tanta la volontà del Cavagliero, che il cavallo habbia da correre, che si spinga avanti la spalla mã[n]ca: ilche fa bruttissimo vedere; & si deve per ogni modo fuggire: & questo tanto più, ch'egli è nocivo, quando si corre all'incontro: perche si fa bersaglio delle parti manche al nemico: si scurta la lancia; oltre à molti errori, ne'quali di quì si può incappare: & il Cavagliero per parer più bello, & piu leggiadro à cavallo, terra diritto il corpo, & non starà agiatamente à seder su'I cavallo: & si volgerà con la visuale, che passi in mezo all'orecchie des cavallo: & stringendo le ginocchia, lascierà cader perpendicularmente, & naturalmente le gambe: & in questo stato le inerbarà, mostrã[n] do tal forza col piede in piano, che la punta d'esso sia al diritto della spalla del cavallo; & che non declini al basso, ne miri ad alto.

p. 187 §2 Et compartirà il Cavagliero la sua carriera in cinque parti; & queste divisioni esso dovrà tener distinte nella sua imaginativa, & nell'operatione saranno collegate insieme in maniera, che i circonstanti non le possino distinguere per tale operatione; cioè; In levare la lancia, in raccorre, in declinare, in ferire, & in riscuotere.

p. 187 §3 Presentato, che si sarà il Cavagliero alla carriera; &

p. 188 §1 postosi bene, levarà primia la lancia dalla coscia: & in quel tempo spronerà il cavallo: & nella prima parte, c'havrà levata la lancia dalla coscia nel nostro modo detto, nella seconda parte poi egli la raccorrà alzando il braccio; il qual formarà angolo: & arrivato che sarà al piano della spalla nella terza parte arresterà calando la lancia: nella quarta poi farà in maniera, che in finire la declinatione colpisca la lancia; nella quinta col detto modo, riscuoterà la lancia sopra la coscia.

p. 188 §2 Potrà ancho prima il Cavagliero armato solo di corazza essercitarsi correndo à piedi: che cosi farà buon polso: & tosto che sarà fatto prattico, montarà sicuramente à cauallo, correndo ò all'Anello, ò al guanto, doppò i quali essercitii tutti potrà mettersi à rompere lancie.

p. 188 §3 Mostrato che s'havrà à i circonstanti di saper correre con ogni ragione, per palesarsi in ciò prattico; & habile, in più modi procederà: come cõ[n] la lancia aperta, & di riverso mettendosi il calce à mano manca: & quã[n]do esso romperà le lancie si potrà talhora fare in aere con la scossa, overo per di dietro al collo; overo in partirsi mettersela in spalla al contrario; & in carriera buttarla ripigliandola nel suo luoco ordinario; overo pigliandola per il calce, & da alto à basso rompendola.

p. 188 §4 Si sogliono ancho rompere le lancie in terra: il quale

p. 189 §1 è modo troppo ordinario, ò in Quintana. La Quintana (per dire ancho questo poco di più) affine, ch'ella habbia più del verisimile; si finge un'huomo fatto di legno posto à cavallo, & sopra un carretto, accioche movendosi, il Cavagliero meglio s'esserciti nel correre all'incontro.

p. 189 §2 Per questo essercitio di Quintana si dovrà piantar la lizza: la quale sarà lunga ducento piedi, & alta cinque: farassi ancho la contralizza, la quale sarà lunga cento cinquanta piedi; & alta da terra due, & mezo; & si farà volta in dentro verso la lizza: accioche non si percuota in essa, quando si allarga no le gambe, per battere il cauallo; cioè; si farà, che la distantia di sopra sia tre piedi, & mezo; & di sotto ne sia quattro.

p. 189 §3 Et affine, che si rompino poche lancie nell'essercitarsi, per far manco spesa, si potrà fare, che il tronco delsa lancia sia lungo quattro piedi, con un cannone in cima ben'acconcio; & fatto in modo, che vi si possa metter dentro con l'incastro l'altro restante della lancia, lungo sei piedi.

p. 189 §4 Sappisi anchora, che quando si corresse à campo aperto, si dovrà avertire di non battere il cauallo con lo sperone manco: accioch'esso non fugga l'urto: & dovrassi stringere il pugno, affine che la resta non dia occasione di stroppiarsi la mano; quando

p. 190 §1 [ave]nisse, che la lancia s'alzasse, & straccorresse per la percossa.

p. 190 §2 Per l'ultimo avertimento dico ancho, che quando si corresse, non per giuoco, ma per far da dovero, col ferro ammollato, si deve sapere, che in arrivando appresso il nemico, s'havrà d'appoggiargli addosso il cavallo per ferire; & questo in linea retta, più che sarà possibile: & si dovrà colpire nel fianco manco; overo si tirarà ad inchiodare la coscia, ch'è parte più disarmata.

p. 190 §3 *IL FINE.*

www.ingramcontent.com/pod-product-compliance
Ingram Content Group UK Ltd.
Pitfield, Milton Keynes, MK11 3LW, UK
UKRC052348051026
15625UKWH00018B/674

9781953683618

9 781953 683618